Computer Integrated Manufacturing
From fundamentals to implementation

Computer Integrated Manufacturing
From fundamentals to implementation

Alan Weatherall
Manager of Advanced CIM Technology
IBM United Kingdom Laboratories Limited

With additional contributors

Butterworths
London Boston Singapore Sydney Toronto Wellington

First published 1988

© **Butterworth & Co. (Publishers) Ltd, 1988**

British Library Cataloguing in Publication Data

Computer integrated manufacturing.
   1. Manufacture. Applications of computer systems
   I. Weatherall, Alan
   670.42'7

   ISBN 0-408-00733-8

Library of Congress Cataloging in Publication Data

Computer integrated manufacturing: from fundamentals
   to implementation.
      p. cm.
   Bibliography: p.
   Includes index.
   ISBN 0-408-00733-8
   1. Computer integrated manufacturing systems.
   2. Production engineering—Data processing.
   I. Weatherall, Alan.
   TS155.6.C654 1988
   670.42'7—dc19   88-14568

Photoset by Butterworths Litho Preparation Department
Printed in Great Britain by Butler and Tanner, Frome, Somerset

Preface

CIM is a commonly-used abbreviation for Computer Integrated Manufacturing. It deals with the fundamental effect on manufacturing industry of integrating manufacturing activities and facilities using computers. The continual changes in business requirements and in the technology of computing make this integration inevitable. It will present major business and technical challenges.

CIM is generally agreed to be a fundamental strategic issue that will effect everyone in a manufacturing company. Some observers see CIM as providing an opportunity for the Western nations to recapture jobs and wealth from the Far East. CIM is therefore very important.

This book is based on a course in CIM which is part of the Production Engineering Tripos for postgraduate-level students at Cambridge University. The CIM course was developed while the author was on a teaching and research secondment from IBM. Since the students are taught other aspects related to CIM throughout the year, including associated project work, the book includes chapters by full-time lecturers from the Production Engineering Tripos and by experts from industry who have relevant teaching experience. Hopefully, the mixed industrial and academic background of the contributors gives the book a balance appropriate to a pragmatic and complex subject.

People who earn their living and the nation's wealth in the day-to-day pressures of manufacturing industry need to develop expertise in CIM quickly and effectively. Study of this book should enable postgraduate students and professional engineers to deal confidently with the subject and use CIM techniques profitably. The book covers CIM and manufacturing systems at a technical level, from description of the conventional 'islands of computerization' to the components of CIM architecture. The business objectives of CIM are described, from analysis of the business environment to cost justification and implementation of CIM systems. CIM is seen as a business tool and not as an end in itself. Each individual and company needs to adapt the tools described in this book to best effect.

The book, like the course at Cambridge University, is intended to give a thorough coverage of a difficult subject, and to communicate principles as well as something of current practice. This should give a firm basis of knowledge in CIM, and develop an understanding that will be valid for many years in changing business and manufacturing environments.

A.W.

Acknowledgements

Many debts of gratitude are owed for the preparation of this book. The development of the ideas contained in it comes from numerous discussions over the years with business and academic colleagues. The Production Engineering students at Cambridge responded to the subject with interest and challenge — an essential further stimulus.

My thanks are due to IBM for giving me the opportunity to return to Cambridge for two and a half years in order to teach and undertake some limited research. This enabled (and obliged) me to analyse such knowledge as I had acquired about manufacturing systems during 20 years in and around the industry, and put it into order. Since the end of that secondment, I have been able to return each year to teach the subject of CIM (with Bill Baillie, the author of Chapter 5), updating the course each year based on current experience.

However, my writing for this book has been done in my own time. The book depends therefore on the contributions of the other authors, to whom I am grateful not only for their chapters, but also for the discussions as to how best to structure this extensive subject within meaningful chapters of a sensibly sized textbook.

Apart from the contributing authors, many friends and colleagues have made detailed comments on draft versions of various chapters, including particularly Frank Ainscow, Colin Andrew, Bill Carter, Richard Hisom, Martin King, Brian Palmer, Chris Strain and John Westley.

Finally, for over a year, my family allowed me to ignore my duties at home so that I could escape to the word processor. Inevitably, most of the consequences of this neglect have fallen on my wife, Coral, whose understanding and unfailing support (except for the occasion of the bent spoon) have enabled the book to happen.

Contributors

A Weatherall

Alan Weatherall read Mechanical Sciences at Cambridge University and subsequently took an MPhil. in Operational Research and Management Studies at Imperial College, London. After working as a consultant and in factory management, he joined IBM in the UK, with technical and marketing responsibilities in database and production control systems for manufacturing companies. From 1982 to 1984 he went to Cambridge on secondment from IBM. He returned to IBM in the UK Development Laboratory, working on IBM academic initiatives in CIM and on advanced applications of CIM. He is a Fellow of the Institution of Mechanical Engineers, and of the British Production and Inventory Control Society.

M J Gregory *(Chapter 3)*

Mike Gregory took a degree in Engineering at Southampton University, followed by a postgraduate manufacturing programme at Cambridge. After some years in industry he returned to Cambridge and has been involved in the establishment of the Manufacturing Engineering Group with responsibilities for the development of the Production Engineering Tripos. He is a Fellow of Churchill College and Director of Studies in Manufacturing Engineering.

W H Baillie *(Chapter 5)*

Bill Baillie joined IBM after graduating with an Honours degree in Physics from Glasgow University. Most of his career has been spent in developing, marketing and implementing database-oriented production planning and control systems. More recently, this area of activity has grown to cover the broader applications of CIM. He teaches the CIM course at Cambridge with Alan Weatherall.

N D Thomson *(Chapter 6)*

Norman Thomson took a degree in Mathematics at Cambridge University and a further degree in Research in Literary Statistics at St Andrews University. He then taught mathematics, including a period as Head of Mathematics at Gordonstoun School, before joining IBM in the education department of the UK Laboratories. He then transferred to an internal consultancy group, where he has developed his present extensive experience in simulation and simulation languages. He is an Open University tutor and has published a number of papers on simulation.

Dr E Appleton *(Chapter 8)*
Ernie Appleton took a BSc in Engineering at Salford University, followed by an MSc and a PhD. He then taught Manufacturing Technology at the University of Nottingham and is currently a lecturer in engineering at the University of Cambridge. He is a Fellow of Girton College, Cambridge, and an author of several papers and books on manufacturing technology and automation.

M E Duncan *(Chapter 10, joint)*
Mike Duncan is a director of Cambridge Robotics Ltd. After specializing in various areas of manufacturing technology at Rolls Royce, Bristol and at Cambridge University, he is now a member of the management team at Cambridge Robotics, involved in product development and factory automation products.

Dr D J Williams *(Chapter 10, joint)*
David J Williams is a Lecturer in Manufacturing at the University of Cambridge and a Fellow of Downing College. He has a BSc in Mechanical Engineering from UMIST, a PhD from Cambridge and a wide variety of industrial experience. His current research interests are intelligent materials processing and the integration of factory automation systems. He has published books and over 50 papers in manufacturing processes and factory automation.

P G Stokes *(Chapter 12)*
Peter Stokes graduated in Physics at the University of Cambridge. He has extensive experience of industrial consultancy, latterly as European Director responsible for Operations Management Consulting for Arthur D. Little. Since 1986 he has been Chief Executive of The CIM Institute at Cranfield. He is a Fellow of the Royal Statistical Society, a Fellow for Operational Research and has published several papers on CIM.

Contents

The business approach

Chapter 1

Why CIM is important

Sound business decisions are at the heart of any successful industrial organization. The computers used in manufacturing industry not only contribute to decision making, but also directly control much of the production equipment. Integrating (i.e. connecting) these computers is the next logical stage in the evolution of computer systems; this integration will inevitably make a radical difference to how the company is run and what it can achieve.

The decision maker (at any level) needs to have access to all the data on all the relevant computers, and to use the computers to analyze the data; this must be a basis for better, more profitable, decisions. These decisions will be implemented with great speed and their effect carefully monitored. The effects of alternative decisions can be simulated. CIM offers this — and more.

What is CIM?

CIM is a label for a set of techniques that are making fundamental changes to manufacturing industry. The techniques are developing because of the logical evolution of computing; batch systems were overtaken by on-line systems in the 1970s, when computers became sufficiently powerful and cheap that they could control communication between host computer and terminal. The continued and highly visible fall in the cost of computing has, by the late 1980s, made it practical for wide communication to take place between different types of computers.

The CIM techniques which need to be understood include:

- the base technology to make this integration happen,
- the information processing techniques to use the much more powerful computer systems,
- the management processes necessary to run the business profitably with these new tools.

These techniques are described in this book.

Manufacturing industry is not alone in taking advantage of integrated computers; banks and supermarkets, for instance, are installing wide networks of computers and terminals. There are however two aspects which are specific to manufacturing:

3

- Many different types of computers. There are basic differences in computer hardware and software between the computers which control shop-floor processes and those which run the financial systems of the business; another example is that computer-aided design (CAD) data is stored in a fundamentally different way to Bill of Material data.
- The many different types of data in manufacturing industry. In most manufacturing companies, the combination of numbers of components, product variations, customers, suppliers, work centres, routings etc. is so large a number that the game of chess seems trivial by comparison. All of this data might be used to optimize the business decision process.

Manufacturing in this context includes all businesses that process materials into a product, including what are conventionally known as process and distribution industries. CIM will provide the most profitable way of running these businesses with the technology available for the 1990s.

Two other terms that cover much the same subject matter are Manufacturing Systems Engineering and AMT (Advanced Manufacturing Technology). However, both have an emphasis more on shop-floor operations, which is only a part, albeit an important part, of CIM.

The multiplier effect

CIM has been portrayed as multiplying the benefits of separate computer systems, compared with adding the benefits in a non-integrated environment. CIM technology will impact strategic issues, such as the introduction of new product lines. It will also affect detailed operational issues, such as whether to accept a short lead-time customer order, agree an engineering change or alter a stock reorder point. The challenge to people in manufacturing industry is to understand enough of the technology of CIM to get the best advantage from it, without investing unnecessary time or money. There will be very few standard answers, because all manufacturing businesses are different. The changes in philosophy inherent in 'just in time' production, the precise and professional long-term planning evidenced in the Far Eastern onslaught on Western markets, the zero defect approach to quality, and many other factors, have changed the rules of the manufacturing game. The rapidly changing circumstances of each manufacturing business and the falling costs of computing will continually modify the way CIM technology can best be used.

So how do people, with human passions and fallibilities, fit into this new, computerized, highly controlled 'Big Brother' world? The answer is that people have the same part to play as they do in an operating theatre, a nuclear power station or the operations room of an international bank. People are still what business is about; they need to work as a team to understand and use the new technology. There is nothing reprehensible in applying the latest technology and every last gasp of human skill to doing the job better.

CIM is not an end in itself. The reason for studying and implementing CIM is to raise the productivity of manufacturing industry, which in turn will add to the wealth of society and improve the quality of life. The individual wants to increase his earning power, each nation needs a stronger manufacturing industry to increase wealth and reduce unemployment, and the world as a whole needs to use its scarce resources better. Millions of people are employed in manufacturing industry in Western Europe, US and the Far East, and are generating enormous wealth. A technology that gives both opportunity and threat to this scale of activity deserves serious and detailed study.

The CIM environment

Experience of using computers, spectacularly emphasized by the wide introduction of personal computers, makes users more sophisticated, so that they expect systems with ever-increasing capabilities. There are no indications that the fall in computing costs is coming to an end. Indeed, the recent discoveries of higher temperature superconducting materials may accelerate the downward cost trend.

Computer systems in manufacturing companies are not of course new. The first Bills of Material on (punched card) computer systems were introduced about the mid 1950s. By the mid 1980s, these systems had developed to quite extensive structures, built on hierarchical databases, covering the applications shown in *Figure 5.4*. Costing and finance were assumed to be included plus maybe some text processing. This type of system is described in some detail in Chapter 5; for most companies it must be the starting point for CIM.

The computer applications shown in *Figure 5.4* often corresponded to existing departments in the typical manufacturing company. In some ways, such flowcharts represent the antithesis of CIM, in that they show a one-way flow of data which preserves these separate functions. The philosophy was 'now that we have run the computer program, we can tell the supplier what to deliver' rather than 'Let's give the supplier the best picture we can of what we are likely to need, wait for his feedback, and then incorporate this feedback into our planning process', which is a CIM philosophy. Measures of how well the systems worked, or were suited to a particular business environment, were usually ignored in pre-CIM systems.

The attitude to people in a manufacturing company is also changing rapidly. Use of the word 'hands' to describe workers in the nineteenth century mills indicated what people were then needed for. F W Taylor early in the twentieth century analyzed the mechanical components of a job (and even suggested multiple managers for each worker as a means of controlling properly each aspect of the working life). In the Hawthorne experiments in 1930, industrial philosophy was surprised to discover that output did not vary linearly with the level of lighting in the workshop. However, the continuous growth in most markets after the Second World War did not encourage further introspection about how the industry was motivated. In the sixties and seventies manufacturing industry in many Western countries was in danger of becoming a political rather than a commercial issue, with sharply divided social theories battling for control. It was becoming increasingly clear, though, that a greater contribution was needed than people's 'hands'.

The Japanese have undoubtedly taught Western nations a great deal about effective management of manufacturing industry. In particular, much publicity has been given to the paternalistic nature of some Japanese companies. However, even the largest companies are feeling the strain of maintaining life-time employment, and it may be that a rather harsher philosophy will eventually evolve in response to the remorseless nature of international competition.

CIM is difficult

He had bought a large map representing the sea,
Without the least vestige of land:
And the crew were much pleased when they found it to be
A map they could all understand

Lewis Carroll, *The Hunting of the Snark*

Whether CIM really is difficult needs to be established with some care. Certainly there are significant commercial pressures to argue both sides, which it is tempting to over-simplify into those academics, consultants and computer salesmen who enjoy new technical challenges and those who don't (with most people who actually work in manufacturing industry still rather confused in the crossfire). This book will hopefully enable readers to make up their own minds on the subject. However it is difficult to see where one goes in manufacturing industry without getting involved with integrated computers, so the subjects in this book need to be understood whether they are called CIM or not. The observation that few companies yet claim to have complete CIM systems installed cannot be held as a valid argument that the subject should be ignored.

CIM depends on the ability to link different types of computers together, which is itself difficult. The publicity surrounding each successive exhibition that tenuously demonstrates (after prodigious investment of time and effort) the linking of a CAD system to a machine tool, or the transference of a not-too-complicated design from one CAD system to another, shows that this is a genuine problem. Rarely do such demonstrations investigate details such as reliability, security and response time, nor do they usually publicize the total costs of the installation.

The limited success of such computer applications gives indication of the problems which will arise when these applications have to be made to work reliably for twenty-four hours a day, seven days a week. However, these applications will become more profitable as computers increasingly become part of the culture of manufacturing industry.

Large computer databases are increasingly common but some of the problems of extracting information from them have not been solved. Individual items from large databases can of course easily be available within two or three seconds. However, computer systems have had very limited success, for instance, in identifying the best material to satisfy a particular specification if interpolation or extrapolation of data on the database is required. Neither has it really been demonstrated how to substitute with one computer screen multiple drawings or reports laid out on a desk (though obviously windowing helps, see Chapter 4).

Even when these issues are solved, there is still no general way of defining to a computer whether a calculated production schedule is acceptable, or of asking it to explain why a particular job has been shipped late. Expert systems and/or artificial intelligence (AI) offer potential solutions to these problems, but specific examples of successful implementation are only beginning to emerge. AI (usually held to include expert systems) is often rather too easily presented as the means of tackling CIM, with the implication that computers will get so clever that it will not really be necessary to understand either the problems or the technology. Over-dependence on computers is a danger that has been recognized in other branches of science, and it may be that if CIM has laws then one of the laws (the First Law of CIM?) would be that however clever the CIM system, someone who understands in depth both the problem and the system will make it even cleverer.

Profitable integration of computer systems imposes the requirement that many people in a manufacturing company understand the entire business and production processes. Education in Material Requirements Planning (MRP) or CAD, for instance, used to be regarded as a major exercise, yet these subjects will be just one of many components in education for CIM. Since technology is never ending, it will be difficult to decide where to stop learning about each of these subjects; this book should give good guidance.

Not, of course, that there is a shortage of literature on the subject. On the contrary, the flow of books and articles on CIM is quite astonishing. However, much of this literature is on individual parts of CIM, and is sometimes rather difficult to put into context. A good example of this is the subject of computer networking on the shop floor, the technology of which can be discussed to the nth degree without any serious reference to what the system is for, and what business benefits will accrue to the company. This may lead to a second law of CIM, that the more expert a person becomes in a single aspect of CIM, the less easy it is to have a proper perspective on the whole subject. What is needed is a framework in which to fit all the available information.

Perhaps the final complexity is that this whole technology is changing very quickly — and it is futile to ask for the roundabout to be slowed down!

There is a distinct consolation for technical and mathematical minds (to revert to Lewis Carroll, who was a mathematician) in the midst of these complexities. CIM systems are likely to be more in tune with their thinking than the 'guesstimated' calculations produced by numerous clerks, which were the sole means of organizing factories until the arrival of the computer. In future, large investments and strategic decisions about CIM will often stake the survival of the company upon understanding and taking full advantage of technological developments.

The conclusion there is that the subject of CIM is difficult and does have a body of knowledge which needs to be learnt in order to deal effectively with it. There is a deluge of literature on the subject which, while confirming the importance of the subject, does not make it easy to get a proper perspective on it. This book lays out the fundamental subjects which are needed, and then builds a structure to implement CIM systems based upon these fundamentals.

CIM is pragmatic

The philosophy of CIM (in spite of the 'laws' above) will always be pragmatic. A generalized total CIM solution is unlikely; businesses will use CIM with the level of integration appropriate to the task in hand.

CIM can perhaps be thought of as a partially explored continent. This book will serve as a map, to help decisions about which part of the continent should be tackled first. Where there are major uncertainties, and the map says 'Here be monsters', fame and perhaps fortune are to be had by slaying these monsters. It may alternatively be better to get established first in the well charted and safe areas.

The expedition to this new continent has to be prepared with great care (and some considerable expense). A large education programme may be necessary to convince the troops and your financial sponsors that you are not going to fall off the end of the world. People with specialist skills will need to be available, either taken on or trained. This book will provide guidance.

Pressure of competition may force the business into parts of CIM that would rather be avoided. It may be that the increasing complexity of technical problems, such as scheduling the capacity of a flexible factory, will oblige another attempt for computer solutions where earlier attempts to computerize have been unsuccessful. But perhaps, since the last time you looked at the map, more parts of the continent have been explored and made safe?

It is important not to be too cautious by waiting until others are established before setting sail. People may not always wish to pass on the benefits of the risks they have taken and the work they have done, so secondhand stories of 'This is the best way' may not be entirely reliable.

A simple search to identify the key characteristics of manufacturing companies so that rules can be developed for CIM implementation (e.g. to determine which system should be installed next by the size of company and number of products) is likely to be as successful as were Sherlock Holmes' attempts to identify criminals by the bumps on their heads. More flexible analysis will be required, as described in Chapter 3.

Simplify the business processes

Part of a pragmatic approach to CIM will be to take the opportunity to simplify the processes by which the business is run, as part of the CIM implementation. This will of course make it easier to install CIM, but it is likely also to give other benefits in terms of clearer objectives for the business.

CIM is competitive

This book abstracts a general foundation of knowledge from developments most often seen in products which are both commercial and competitive. Detailed reference to such products, and any comparison between them, should be avoided as far as possible in a textbook of this type, and would in any case become quickly out of date. Some subjects, such as database design or 'compatible' computers, can scarcely be broached without invoking major commercial sensitivities. More subtly, an intellectual analysis of possible future developments could appear to be an implied criticism of a current product. Even anticipation of international standards might be held to have a commercial angle. Every effort has been made to avoid any solecisms.

Structure of the book

The book is divided into four main sections. The business approach to CIM is covered in the first section, since it is fundamental to CIM that it provide solutions to real business problems. In the second section, computer components and systems are described under the heading of 'islands of computerization', covering what might be termed pre-CIM systems. The components of a CIM architecture introduce the third section, followed by chapters on the applications of CIM in manufacturing industry. The fourth section covers implementation, including discussion of some of the major initiatives in CIM which are currently taking place. Some tutorial questions are included to assist further study of the subject. An index is supplied at the end of the book.

Business approach to CIM

In order to use CIM to its full effect, the opportunities and problems of a business must be analyzed in terms of its objectives and of the business environment. In Chapter 2 techniques are presented for doing this and for agreeing the financial benefits of CIM. A set of problems can be defined that are common to many manufacturing companies. The demands that CIM will make on the people in the business are explored and the importance of quality is emphasized.

There will be as many different implementations of CIM as there are manufacturing companies. A technique which gives good benefits in one company may be of little value in another company, or even in the same company when the business environment changes. It is therefore important to analyze different types of manufacturing organizations as an aid to auditing system effectiveness; this is covered in Chapter 3.

Islands of computerization

In Chapter 4 the basic hardware of computers, including shop floor hardware, is described, and the relationship between operating systems, programming languages and databases. Hierarchical and relational databases are covered, with the communications systems needed to access and control this data. The functions of the information systems department in a manufacturing company are discussed.

Chapter 5 describes the flow of information in a manufacturing company. The conventional operation of computer systems in manufacturing industry is reviewed in a structured manner. This leads to a description of the 'islands of computerization', including master production schedule planning, material requirements planning, and capacity planning.

Computer-based simulation has long been a useful tool in manufacturing companies and is described in Chapter 6. There are many simulation systems available; typical features of such systems are described and categorized. Practical limitations of the technique, including the statistical and analytical skills needed, are discussed. Linear (or mathematical) programming is included.

It will be different with CIM

Chapter 7 identifies the components of an architecture (or framework) for CIM; integration is required at different levels. Hardware and system software will need to interface, based on international standards. Applications also need to integrate, i.e. the customer order system must be able to communicate with the stock system, which may not be easy even if the two systems are on the same computer. Decision-support systems, including expert (knowledge-based) systems are described.

Chapter 8 covers product and process design for CIM. The business requirements described in Chapter 2 impose new demands on the designer, whilst the availability of CIM technology provides massive new opportunities. Techniques for categorizing manufacturing processes and products can be used as input to a CIM-based design function.

Planning and scheduling in a CIM environment are described in Chapter 9. There will be fewer people in a production facility to take detailed planning decisions, and indeed the automated processes will make it increasingly unlikely that human planners will be able to handle all the data and react quickly enough.

Chapter 10 reviews the forms of advanced manufacturing technology. It covers the application of machine centres and robots, the design and use of more flexible cutting systems, cells for factory automation and the practical use of industrial machine vision systems. Future technologies are likely to include increasing use of sensors and applications of artificial intelligence techniques on the shop floor.

The integration of computer systems between suppliers and customers under CIM is explored in Chapter 11. The apparently simple idea of sending quantities of component requirements to a printer installed in a supplier can develop rapidly into a complex and extensive communication system.

Wouldn't it be safer to wait?

Profitable implementation of CIM is the ultimate objective of the exercise; implementation of CIM systems is covered in Chapter 13. There will be extensive requirements for education and training. Effective project management will be vital, as will commitment to quality of product and business processes. It is important to measure benefits to ensure that every scrap of the advantages of CIM are taken advantage of in a fully professional manner.

Chapter 13 explores possible lines of further study, where CIM might be going in the next few years, and what might be beyond CIM (Meta-CIM perhaps?). CIM is the subject of major governmental and industrial initiatives, including the ESPRIT CIM architecture project and major educational programmes; such projects emphasize the importance and relevance of CIM.

Tutorials

The tutorials contain a set of questions to prompt further study and to help ensure that the ideas of the book have been understood.

Reading the book

The book has been planned so that readers with different perspectives and experience can take different paths through it. Chapters 4, 5 and 6, although in a logical sequence, are intended to be stand-alone; they will normally be taken on a first reading of the book but, for those who wish, can be left until later. Chapter 7 sets the framework for the following four chapters, which are basically in a sequence of design, plan, make and ship, but can also be read as required. Some readers may even prefer to begin at Chapter 12 and work backwards from the implementation plan!

However, only those who are already running their own manufacturing company should neglect Chapters 2 and 3; as has already been said, and is likely to be repeated, CIM is not just about computing techniques, it is about achieving business results.

Chapter 2

Business perspectives for CIM

From business objectives to CIM solutions

To understand how a business will benefit from CIM, the opportunities and problems must be analyzed. To do this (whether or not the acronym CIM is used) the objectives of the business have to be agreed. Strategies have to be devised to satisfy these objectives and overcome the business' problems; the resultant plan must gain the commitment of the people involved. Because of the fierce and continual nature of international competition, some problems are common to almost all manufacturing companies. By translating these problems into the system requirements, we can define what is needed from CIM. This approach to CIM logically establishes the value of CIM to manufacturing industry.

Each company which implements CIM (or parts of CIM) will have its own unique systems. Understanding the new technologies within CIM requires considerable education — for specialists to define and implement the new systems, for the people who will directly use the systems and for the entire work community.

What is manufacturing?

As discussed in Chapter 1, manufacturing industry in this book includes electrical and electronic as well as conventional metal manufacture. Chapters 5, 8, 9, 10 and 11 will show that the whole business process is encompassed by CIM, from conception of the product through to shipment and after-sales service. Hence the CIM approach is also applicable to companies whose prime business is the distribution of goods. Computer integration is a technological evolution that cannot be ignored any more than can the on-line computer systems that preceded it; one of the results of this integration is to intimately connect parts of the business that would previously have been considered to be separate.

Objectives of a manufacturing business

Most manufacturing businesses (and indeed most other organizations) have a basic objective of long-term survival. However, to stay in business, more immediate objectives are needed which must be specific and must be measurable. Typical examples include:

maintain good relationships with customers,
take a significant market share,
grow with the market,
have a presence in major industrial countries,
earn profit,
introduce new products ahead of competition,
be a low cost producer,
ship high quality products,
be a technological leader,
maintain a well motivated work force,
contribute to the community,
be independent of any single customer.

It is clear that such objectives are only meaningful if the business is sound and able to publish acceptable financial figures. These figures must be satisfactory to maintain the interest of shareholders, banks, suppliers and customers, and also to pay next weeks wages. However, the focus in this chapter will be on objectives which are specific to manufacturing and can be related to CIM. The financial facts of life will be taken as read.

Lack of clear objectives about market share and technological leadership may well be a symptom of a manufacturing company which has abandoned any real hope of long term survival.

The actual measure of the above objectives requires tight definition, particularly the non-financial objectives. There are many aspects to be considered.

Market share

For many companies, market share is a vital measure of how the business is progressing. This is familiar ground in the automotive industry, for instance, where the figures get significant press coverage. The benefits of large-scale operation, and of accumulated experience in product development, manufacture and marketing are now well understood within the concepts of experience curves. For those companies which face international competition (or might do at some time in the future) share of the world market needs to be closely monitored.

It is commonly agreed that the conditions of aggressive international competition which have recently emerged have not generally been seen since the nineteenth century. The oil crises of the 1970s and the high growth rate of the 'Pacific Basin' economies, among other factors, have contributed to this situation. The three decades of relatively steady growth for Western industry which followed the Second World War will not be repeated in the foreseeable future.

The image of an army with no respect for national boundaries, and choosing when to concentrate its strength on key targets, is very relevant in the competitive state of manufacturing industry today. As in war, attempts are made by Governments to agree on conventions (called quotas) and these are occasionally followed. The reality of the battlefield (where conventions may be overlooked) can be likened to the consequences of free customer choice (which may override quota agreements); victory is likely to go to those who are the best led, the strongest in engineering weaponry and the most fully committed. A company which does not understand the strategy of its competitors and does not maintain a viable market share is unlikely to survive for long.

It is also beginning to be widely understood that losing the world trade war in manufactured goods has consequences for a community which may not be any more acceptable, in terms of social disruption and financial deprivation, than some of the effects of conventional war.

Technology leadership

Technology leadership requires choices to be made as to which technologies are central to the business. Numerical control (NC) can be used as an example. If NC is a key competitive component of the business, it may be necessary to keep both expertise and equipment in-house. On the other hand, a company could take the viewpoint that competition between NC suppliers will ensure a high quality, leading edge and low-cost service. This will have the advantage of avoiding some major investment decisions and the problems of complex equipment maintenance. Between these two alternatives, it may be possible to keep only the expertise in-house in order to negotiate better with the suppliers, taking due care to ensure that the in-house experts do not become out of date. Each company will have its own decisions to take about which technologies are vital to the business, but in any event CIM will be a major issue of technological leadership for most manufacturing companies.

Product and process quality

Increasing quality of the product is now a major objective for all manufacturing businesses. High quality is forced upon a company that wishes to export widely, and hence much of the initiative in the drive for quality has come from Japan and the Far East. Few products will make money if their quality is so bad that free maintenance work must be carried out under guarantee immediately after shipping them half way round the world. Quality of business processes (e.g. simplified systems, up-to-date facilities, sufficient training) is an important component of product quality. There are many procedures for grading and monitoring a company's quality programmes (e.g. British Standard 5750).

Low-cost products

Unfortunately high quality does not of itself represent a justification for a high product cost. High quality and yet low-cost electronic goods, for instance, can be seen in every high-street. No company these days has an excuse for not knowing exactly what costs are, and for not being able to show real activity to drive these costs down.

Even for contracts with Government agencies the world is changing; there is a move away from the 'cost-plus' method of charging. In the world of cost-plus, the supplier to the Government adds a profit margin to the costs which have been incurred, and presents the resulting figure on the invoice. In theory, the higher the cost, the higher the profit. Although the cost-plus system has the major benefit of allowing Government subsidy of selected companies without a public admission thereof, the system is wasteful in discouraging innovation and cost saving exercises.

Computer systems, particularly when CIM is introduced, will be instrumental in improving this situation by removing the excuse that detailed control of the costs of the business is impossible.

Good employer

It is also a serious business objective is to be seen to operate as a 'good employer'. There is of course specific legislation in some areas (e.g. sex and race equality, dismissal procedures, safety and in some countries age discrimination) and even the most hide-bound employers are moving away from multiple levels of lunch time cafeteria. However, the introduction of new technologies such as CIM and the need for frequent re-training are themselves a major driving force for a more flexible approach to dealing with people. The fork-life truck driver may well have the potential to be the new computer programmer, and apart from other considerations, it may save a lot of money to find this out before incurring the cost of advertising. Hence there are real pressures to ensure that people are treated in a manner which brings the best out of them, whilst necessarily staying within the bounds of what is practical for the business.

Community involvement

It should also be a serious business objective to be seen as a good corporate citizen and member of the community. Apart from the occasional company sponsorship of suitable artistic functions, manufacturing industry as a whole needs to continue conscious efforts to make the general public aware of its contribution to the community. Such a contribution was not seriously questioned a few years ago, when a much higher percentage of the population worked, or knew people who worked, in manufacturing industry. The decline of jobs in manufacturing, although a world-wide phenomenon, has lead to an atmosphere where it is easy to knock the industry as being a cause of unemployment. This does not put the industry in a good public light when questions of planning permission or local taxes are being discussed.

Profit

Profit is still an emotive word, representing to most people a simple measure of the contribution of a (legitimate) business to the community, and also the means of securing funds for future developments. However, to some people such objectives, particularly those related to cutting costs, will seem to be unimportant — if not positively anti-social. But whatever measures are accepted as valid for the performance of individual manufacturing companies, there is a world-wide need for more manufactured goods, ranging from simple agricultural tools to advanced medical equipment. Hence, improving the performance of manufacturing industry must on any grounds be accepted as a valid objective.

Customer relationships

This is currently a fashionable objective, again lead by businesses in the Far East. The success of customer relationships is hard to define, unless it be by the conventional measures such as market share and profit (although surveys of customer satisfaction will have a value for larger organizations). Hence identifying this as a business objective may be a psychological decision (to encourage the troops) or an expression of company philosophy.

Within the context of the appropriate business and social objectives (including a high standard of business ethics, which is an unwritten component of all the above objectives), the problems and opportunities for a manufacturing company can now be analyzed. This is done to demonstrate the importance of the contribution which will be made by CIM.

Identifying business opportunities and problems

To identify the opportunities and problems of a business requires a structured approach to planning. Two different approaches are often familiarly referred to as 'top-down' or 'bottom-up'. The top-down approach tends to start with external factors, for example that the company must get into the South American market within the next 5 years, or (more urgently) that the company lost money last month. The bottom-up approach looks at more technical and detailed problems.

Whatever the approach, a major target must be to simplify the business. There are, of course, many situations where simplification of systems and products will produce major business benefits without the installation of new computers. This simplification should precede, and not preclude, consideration of the business benefits of computer systems. The approach to business systems architecture which has been developed within the ESPRIT programme is discussed in Chapter 13.

In some organizations there can almost be a pride in enduring overly complex procedures (to which it should be added that complex procedures have the additional benefit of keeping in their place newcomers to the department and other outsiders). A further defect in a business which can often be seen as a virtue is a lack of proper organizational support for the job. Both of these — complexity and inadequate support — create job satisfaction because they create visible tasks. Filling in a complex (but unnecessary) form, or handwriting a document that should have been prepared on a word processor, is an easier task than the often less tangible activities that professionals and managers should be tackling. Furthermore, everybody is seen to be working when they are doing the business equivalent of white-washing coal.

It is therefore important to adopt a somewhat coldly analytical attitude to the existing operation of the business, and not to accept it at its own valuation or take it too seriously; departments can actually be disbanded (or combined), their objectives or means of operation can, and sometimes should, be radically changed. And to repeat, simple straight-forward solutions are sometimes (but not always!) the best.

The bottom-up approach

To many of the people working in a manufacturing organization there is a need for bottom-up planning to deal with every day internal problems, and to take advantage of opportunities to improve productivity. Questions such as 'Why are these parts always late?', 'Why do engineering changes take so long?' or 'Why do we change delivery dates for our suppliers so often?' are hardy perennials, which many people will recognize.

These questions will always arise — even with the most complete CIM systems — since there will alway be frustration at anything short of perfection. A delay of a few seconds in a telephone being answered causes irritation which is not to be assuaged by the thought that the pigeon post would take rather longer.

What is needed therefore is to be able to convince people continuously that, within the limits of commercial viability, maximum effort is being made to put their job knowledge and experience to good use. One method is the introduction of what are commonly called quality circles. These usually have a wider span of responsibility than product quality, the word quality being used to refer to the quality of the whole business process. Quality circles consist of small groups, say six to eight people, who meet regularly, perhaps every week or fortnight, to solve specific problems which the members of the circle have usually defined themselves. Leadership of such groups is normally independent of line management organization, though facilities and training are usually provided on a formal basis. It is a matter of some incredulity that quality circles are not a universal feature of manufacturing companies (and most other business organizations), since they offer obvious potential for involving a wide range of people in responsibility for their part of the organization. Quality circles were first used extensively in Japan (where the very high standards of education may have made them easier to implement).

Suggestion schemes are another proven way of reducing costs and increasing the involvement of people in the business. Absence of an effective suggestion scheme in almost any organization would seem to indicate a management that is not really trying to do an effective job. Even the simplest of suggestions can save the organization a great deal of money.

However not all problems can be solved bottom-up. Wider problems may require major restructuring of the organization; changes in one department may be required to reduce costs in another. Some problems require changes in working practices that are genuinely difficult to implement. For example, introduction of a CAD system requires increased skill in the Design Office, both in selecting and using the new equipment. The business benefits of the new CAD system are not necessarily totally convincing to the individuals who are required to put their careers on the line in order to implement the new technology. Similarly, an on-line system for stock control has obvious benefits to the rest of the organization, but it can be a lot to ask the storekeepers to instantly introduce the degree of accuracy conventionally achieved in a bank, and do this within the spotlight of public knowledge whenever a figure is incorrect. Furthermore someone has to decide whether the cost of these new systems is justified by the often immeasurable business benefits which include such words as 'quicker' and 'better'. Hence valuable as the quality circle and suggestion scheme types of approach are, there is often a need for a top-down viewpoint and a more structured approach to evaluating business benefits.

The top-down approach

One top-down approach is to take a small team of key people and give them a joint responsibility for defining the problems precisely and then agreeing the solutions. A task force is one of various names for such a team. Office facilities should be provided for the team and an atmosphere of confidentiality and trust must be established. Shortened versions of this approach can be run successfully, but ideally a number of stages should be gone through, including problem definition, problem prioritization, interviews of key people, and action planning.

As a means of problem definition, the team could jointly agree single sentence statements of the problems, such as:

'we do not know what our stock levels are',
'receiving often do not get the paperwork about purchase order changes in
    time',
'we do not get enough quotations from potential new suppliers',
'there are too many late deliveries by suppliers',
'the different Bills of Material do not agree',
'the cost reporting system is inaccurate',
'we need to reduce the capital tied up in raw materials and inventories',
'queues of work in the machine shop are too long'.

It might be thought that a group of people who know the company well could go on describing such problems forever, but eventually most points have been covered and the list is agreed to be more or less complete. As a crude but comprehensible measure of priority, identifying the business area with the longest list of problems may be useful. A more meaningful approach is to ask the appropriate people to estimate the value to the business of solving these problems. It may be felt that such detailed analysis is not necessary because there is a common agreement about the main problem (e.g. that the company does not offer enough options with its products), but it is not always certain what is cause and what is effect. By a variety of means, out of such analysis should come a short list of the major business problems to which a high priority should be given.

At an appropriate stage, the team could be split into units of two or three who would formally interview key personnel in the target business areas about their perception of the scope of problems and opportunities. Three people is probably ideal, allowing two to ask questions and one to record the conversation. This record should be formally (and confidentially) written up and signed by those involved. This will reassure the interviewee that his points have been understood, and put a professional discipline on the investigating team. The problem specifications, and the results of the interviews should then be synthesized to produce a formal statement of the problem and the required solution. Individual confidentiality will thus be preserved.

A typical summary statement at this stage might be as shown in *Table 2.1.*

Table 2.1 Major problem statement

Major problem: organization of engineering changes

Problem description
- There are too many E/Cs
- E/Cs take too long to get approved
- Delays in the E/C cycle cost money in avoidable rework

Causes
- The paperwork system is cumbersome
- Difficulties of communication between design departments and the manufacturing plants in different locations
- The costing system encourages one-off fixes rather than proper investigation and permanent solutions

Outline solution
- Set up a database of E/Cs with access from Design and the Plants
- Introduce a computer-based message system for progressing E/Cs and providing the relevant information
- Produce control reports on the cycle time of E/Cs
- Give objectives to a senior professional for cycle time reduction

Dependencies
- Strategy statement to be agreed by the Board
- Cost justification to be established

Such a problem statement has the authority of a structured investigation, which will hopefully lead to its general acceptance even by those who would have given the higher priority to other applications, such as links with suppliers or a shop floor Local Area Network (see later in the book for more details on these two subjects).

Consensus does not of course necessarily produce the best outcome for the long-term health of the business. It may be necessary for someone with a wider strategic view to override the results of this type of analysis. But at least the position is clear and responsibility is acknowledged.

Primary business problems

Clearly each organization has some unique problems; most people will be able to describe the problems as they see them in their own organization. Given the dominance of world-wide competition and high rate of technological change, there will be some problems which most companies are likely to have in common.

A prime problem common to many companies is the need to reduce lead time for the introduction of new products. There may be an increased demand for product 'facelifts' or for short special batches for major customers. Minor product variants are usually required for different national markets, even if these only consist of different languages on the labels.

The need to improve product quality, perhaps because competition is offering extended guarantee periods, is another common problem. The solution may involve more fundamental changes than a few extra quality control people on the shop floor and in Goods Receiving. Component or even product redesign may be necessary to produce significant improvements in quality.

Similarly, cost reduction can require basic changes to products and production processes as well as (or even instead of) bigger sticks and more carrots for the people in the factory. Indeed, it might be held that much of the well publicized comment blaming the shop floor for cost problems (*vis à vis* the Far Eastern manufacturer) obscures a lack of technical capability to make the engineering improvements which are really required.

The need to respond to swift and unpredictable market changes requires flexibility in production processes and systems; there are real limits to the speed with which new production systems can be set up. Capital machinery will often be typically on a six to nine months lead time. New computer systems for production control often require six to twelve months to install by virtue of a development cycle which usually includes consideration of other business priorities (as discussed above) and the need for wide education programmes. These factors are illustrated in *Figure 2.1*.

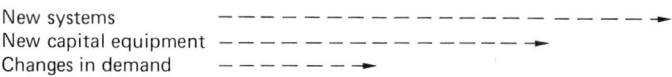

Figure 2.1 Relative timescales for change

If production levels are lower than planned, emphasis within Production Control will usually be on controlling material to keep work in process to a minimum. If production is to be increased to the maximum level possible then capacity control rather than material control is the top priority. These two different strategic requirements may require different production control systems.

A survey of primary business issues in manufacturing industry by INSEAD in 1985 identified similar factors. *Table 2.2* shows competitive manufacturing priorities, with some interesting differences between Japan and Western industry.

The purpose of quoting this analysis is not to place too much reliance on the details, although it is tempting to suggest from *Table 2.2* that Europe and North America are just catching up with the Japanese emphasis on quality of the last two decades, and that the Japanese are now onto the next problems. However, even though each company is unique and the ranking of its problems likely to vary from month to month, it would seem safe to conclude that primary and common business problems do exist.

It will be seen that, within the context of looking for simple solutions, many of the changes that are needed in manufacturing companies to solve these primary business problems can be provided by computer integration.

Table 2.2 Competitive manufacturing priorities

European ranking		North American ranking	Japanese ranking
1	Consistent quality	1	*
2	High-performance products	*	3
3	Low prices	4	1
4	Fast deliveries	3	6
5	Rapid design changes	6	2
6	After-sales service	5	7
7	Rapid volume changes	7	5

* Not included in first seven factors. Table adapted from DeMeyer, A. (1986) Manufacturing strategies in Europe compared with North America and Japan, *Managing Advanced Manufacturing Technology*, ed. C. A. Voss, pp. 19–28. IFS Publications

This will lead to a clearer understanding of what CIM means and how CIM is logically the next stage in the evolution of computer systems in manufacturing industry.

The business characteristics of CIM systems

CIM can now be introduced in relationship to the primary business opportunities and problems discussed above. This will serve as an introduction to the more detailed studies in the later chapters (and the following summary begs some serious technical questions that need to be understood if effective CIM systems are to be implemented). By its nature CIM crosses conventionally accepted boundaries of company departments (and incidentally of academic study). Hence a description of CIM cannot start at the beginning and proceed to the end, but must appear to jump from one subject to another (and then perhaps back again).

By filling the organizational gaps which create the primary problems, CIM can be seen logically to tackle the issues shown in *Table 2.2*. As described in Chapter 1, CIM is not an easy subject. The architecture (or framework) for these system functions will be described in Chapter 7.

The design function

The first conceptual design may well consist of a few sketches on paper. Depending whether the design is of a simple component or of an aeroplane, there will be different stages between the conceptual design and the start of detail design. In a growing number of cases, the designer uses a solid modeller early in the design cycle (see Chapter 8 for more detailed discussion of this function), thus being able to automatically create conventional component drawings. Computer packages are likely to be used to prove the detailed design, e.g. for strength via finite-element analysis or for logical correctness in electronic circuitry. Information on long lead-time components and critical raw material requirements will be fed via the Purchasing Department to suppliers as early as possible.

Manufacturing design must proceed in parallel, to create tooling, manufacturing instructions and, if appropriate, programs for numerically controlled machine tools and robots. This early involvement of manufacturing people with the design function is very important, and is part of business integration; the process can be summarized as in *Figure 2.2*. The many requirements of designers to access text systems (catalogues, technical reports etc.) are indicated.

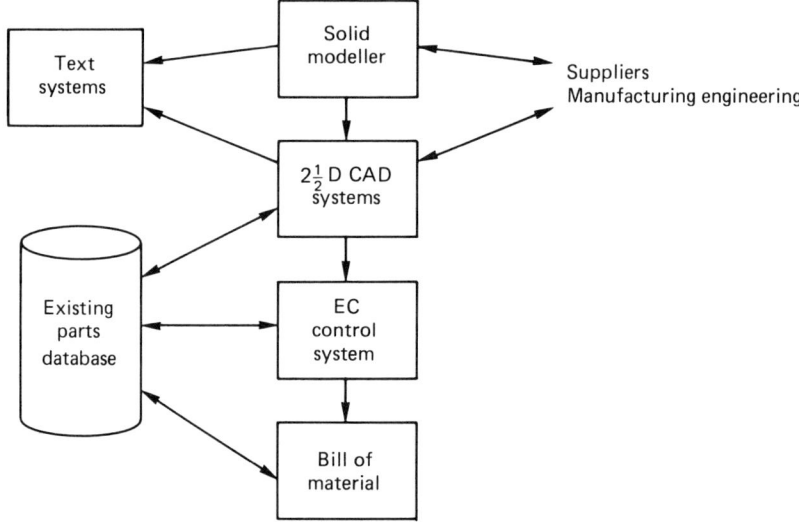

Figure 2.2 The integrated design process

This initial figure indicates the degree of integration with other computer systems; the potential benefits in terms of reduced lead time for new and modified products are apparent.

Engineering change control

In many manufacturing companies, there is a major gap between the CAD System in the design office and the Bill of Material System used for MRP and manufacturing control. Often these two systems are on different computers, and even if the data is held on the same computer, the file structures (i.e. the organization of the data on the computer) are quite different; hence it is not technically possible to combine the two systems into one.

The different interests of the users also make interfacing the CAD and Bill of Material systems an issue of some difficulty. The structure of the product as the designer sees it can be very different to that required by manufacturing, for example. The designer (and the cost accountant) will want a list of all the parts associated with an optional feature of the product, whilst the manufacturing view will be interested in the assembly sequence, with possibly no concern as to what is optional and what is standard in the sales catalogue.

The link between these two systems is effectively controlled by the engineering change system — in many companies still formalized by paper documents that describe details of the change and the reasons why it is being proposed. Such documents are circulated to all interested parties for their comments and ultimately a signature for approval. Timescales of days stretching to weeks may not be unusual for the progress of such documents around an internal mail system. With CIM, this paper document will be replaced by a computer-based message system, using passwords to formalize approval and databases to maintain up-to-date information as indicated in *Figure 2.2*.

However, even with simple changes, there are difficulties in marshalling all the relevant data. An engineering change may depend on other changes, and reference may need to be made to several drawings simultaneously. Access may be needed to other information such as cost data, supplier catalogues, or stock on hand. Hence, a complex system is needed to manage this process and speed up this interface, the definition and implementation of which is a part of CIM, and is discussed further in Chapters 8 and 11.

Customer order planning and control

This involves planning for future materials and capacity, accepting orders against plan and controlling orders in the manufacturing process. A conventional MRP system will calculate the components required for a proposed build schedule, defined on the master production schedule. An MRP system does not usually answer the question 'What can best be built given the current stock levels and components on order?' (which is what many people expect when they are first introduced to MRP). This is an area where improvements would be expected with the integrated system shown in *Figure 2.3*.

Capacity Requirements Planning (CRP) systems will be more effective if integrated with the MRP system and with shop floor systems (for feedback on current production progress). A CIM system which produced an MRP schedule of components required for a set of customer orders, then evaluated whether capacity was available, communicated with suppliers, fed the consequences of capacity limitations back to the MRP system, and then iterated around this loop until the optimum schedule were achieved, would clearly tackle problems of on-time delivery and rapid volume changes (see *Table 2.2*).

Planning with CIM

In effect, the computer is still not widely used as a planning tool. Apart from the trivial arithmetic contained in an MRP system, most computers are used in what one might call 'execution mode', i.e. the computer system presents information to someone who will decide what to do next. The data is usually processed only in a simple manner (e.g. Is the stock level below re-order point?, Is the order more than ten days late?) in order to help identify the need for some action. The computer is little more than a

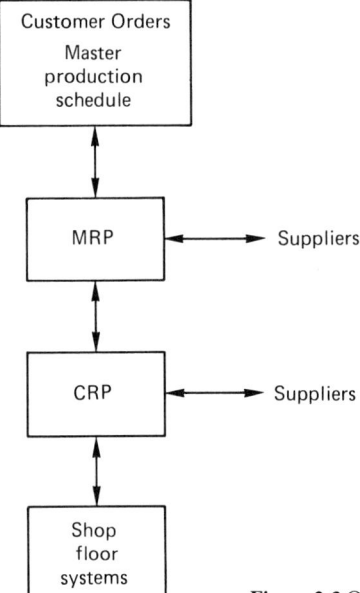

Figure 2.3 Order processing integrated with CRP and suppliers

notebook that can be written into and read by many different people at the same time. CIM systems will need to process the data in a serious manner to determine what the next action should be.

Some of the processing of this data will require innovative work on the software. It is not going to be possible to specify as part of the logic of a computer program exactly what decisions are to be taken in all circumstances: for example, if the special order is three days late, the assembly line has been down for four hours, there has been no overtime so far this month and the Managing Director is due on Monday for a progress review, then overtime will be worked during the weekend. It is rather difficult to define this set of circumstances to a computer program!

The computer programs to handle this sort of situation are called expert systems, within the general subject of artificial intelligence. Their possible use within this context is discussed in Chapters 7 and 9.

Simulation will have an important planning role in CIM systems. The shop floor layout will require simulation if transport is to be automated with automatically-guided vehicles (AGVs) and conveyors — both much less flexible than people. Shop load calculations will become more complex as lead times fall; a lot of companies seem to maintain long lead times because they can't handle short-range capacity planning. Simulation is fully discussed in Chapter 6.

Also described in Chapter 6 is linear programming (LP), another computer-based planning tool that will be of increasing value. Because of the availability of software to prepare the input data, and the cheapness of computer processing power, LP is now of value in wider applications than the scheduling of oil refineries.

Shop floor communication and control

Better communication on the shop floor is another means of reducing lead times. Manufacturing industry has not yet reached the detailed control found in the process industries — with good reason, since there is a much greater mix of work through a typical factory. Paper systems with due dates printed on the job documentation can still be seen on the manufacturing shop floor, making rescheduling effectively impossible. The shortage list therefore becomes the source of priorities for the foreman; high work-in-process and long lead times are the inevitable result. The statistics that show components in machine shops spending ninetyfive per cent of the time in queues are all too well known.

It is clear that computers now have the processing power continually to reschedule the shop floor for optimum performance, doing detailed calculations of actual load on each separate shop floor facility. This will increase flexibility and reduce the money tied up in unnecessary work-in-process. The computer should have instant access to the data on the shop floor, which requires a sophisticated system of automatic data collection. Having calculated a new schedule, the computer must then be able to issue commands to many different items of shop floor equipment and put the new schedule into effect. Industrial Local Area Networks (ILANs) on the shop floor are likely to be required. Selection and implementation of the appropriate ILAN involves some significant technical decisions (explored in Chapter 7) and also substantial improvement in the availability of skills for installing and maintaining electronic and mechanical equipment.

The preparation and testing of programs for shop floor equipment (e.g. robots and NC machine tools) can be a major area for cost and lead-time reduction. Techniques for preparing such programs off-line, thus avoiding interrupting production, are being developed. They are discussed further in Chapter 10.

Integration with suppliers

Computer-based links to suppliers can obviously save time. These links should be two-way (i.e. from customer to supplier and back again) and, as well as the conventional date/quantity information contained in a commercial order, should enable communication of capacity information and CAD-based technical data. Techniques for this, including the development of international standards, are discussed in Chapter 7 and 11. Integration between customers and suppliers (plus the banks for financial transactions) has implications, of course, for financial planning and control.

Product costing

New approaches to product costing may be required (e.g. spreadsheet packages are very suitable as a calculation tool). Conventional calculations allocate overhead charges to direct labour hours. The distortions produced by this system, derived from charging methods introduced for Government contracts during the First World War, have possibly played a non-trivial part in the decline of Western manufacturing industry. Be that as it may, applying conventional overheads to direct labour hours becomes irrelevant as the number of direct operators decreases. For major CIM projects, product costing could be a factor in the justification process, and in this context is discussed further below.

In-house publishing

Integration of computer systems in order to access textual data is another important area of CIM. Technical descriptions and suppliers' catalogues are ideal for computer systems that can search for 'key words' in large volumes of data very swiftly. However there are real difficulties both in describing exactly what data is needed and in accessing multiple sources of data simultaneously on a single computer terminal. The computer really has not helped much if the user has to search through computer displays just as through the pages of a book.

The computer can also create textual documentation, working directly from the CAD and Bill of Material systems to high quality printers. This is described in Chapter 11.

Hence, it can be seen that CIM systems in principle do focus on the primary business problems. Before going into more detail in the following chapters, techniques for agreeing the value of CIM systems should be understood.

Quantifying the benefits

At some stage it will be necessary to grasp the nettle of assigning numeric values to business benefits. Establishing collective agreement can be difficult, partly because the numbers are naturally indeterminate (e.g. what would be the percentage increase in sales if product lead times were reduced by two weeks?). Also people will naturally use the process of quantifying benefits to support their ideas about what the problems really are. For instance, someone who genuinely believes that most of the problems of the business stem from lack of detailed capacity planning in the machine shop is going to produce very different cost benefit numbers from someone who thinks that most of the company's business problems would go away if only someone would discipline the suppliers to deliver on time. This sort of conflict can be substantially eased if there is a forum for open discussion and comparison of these perspectives, as discussed above.

A useful technique to help establish meaningful numbers for benefits is to split the gross figures into different categories, good examples of which are:

- cost reduction,
- cost avoidance,
- increased sales revenue.

Thus cost reductions, which derive for example from reduced maintenance expense following a quality programme, can be separated from the less definite cost avoidance of a future price increase. This is reasonable, since cost reduction figures are bounded by the maximum figure for existing costs, while cost avoidance figures have no such limit. For instance, in the year after a computerized purchasing system is installed, it is not possible to establish what the average increase in suppliers' prices would have been without the new system. If the actual increase were 5% on average, then a claim of 10% would perhaps justify the new system (i.e. the new system would be credited with a benefit figure equal to 5% of the purchase spend). However, a claim of 15% or 20% could be made just as easily, doubling or trebling the apparent benefits of the system.

Figures for increases in sales revenue can also be kept in a category of their own, if only because such numbers tend to outweigh cost reduction and cost avoidance figures (almost any expenditure can be justified by the claim that it will give a 5% increase in sales!).

Weightings can then be given to these numbers e.g. 80% of the cost reduction figures (few things in this world are certain), 40% of the cost avoidance figures and 10% of the increased sales revenue figures. A simple numeric illustration of this is shown in *Table 2.3*.

Table 2.3 Weighted cost benefit calculation

Category	Base value	% applied	Weighted value
Cost reduction	150K	80%	120K
Cost avoidance	450K	40%	180K
Increased revenue	1100K	10%	110K
Total accepted value of cost benefit			410K

Not all the problems and solutions will have an immediate relationship to computer systems. For instance, computers cannot directly help with a shortage of skilled machine shop personnel, or uncertainties in the international market for a raw material. However, there will be few problems, including the two just mentioned, which cannot to some extent be improved by more detailed computerized planning and control. The benefits need to be assessed.

It is always worth while to ask the question 'What other business benefits can we achieve from it?' For example, a new stock control system may be justified by the consequent reduced stock levels, but the new system may also give the Service Department the opportunity to set up a new type of maintenance contract. The absence of this new business had not figured as a problem, but the revenue from it would undoubtedly be a benefit.

Hence numbers can be defined for the annual benefits of the proposed CIM system (or by some other unit of time). Lack of a proper analysis of the benefits may indicate that the proposed system solution has not been sufficiently carefully considered; it should perhaps be treated with caution, since there may be further problems and obstacles to implementation that have not been identified. Quantifying the benefits is a necessary prerequisite to the justification of whatever expenditure is required.

Justifying the investment

It is usually necessary to provide an arithmetic 'proof' that the investment required for any new system is justified. This involves:

- specifying how much the next stage of the CIM system will cost,
- agreeing the benefits (as described above)
- doing some arithmetic to establish that the scale of benefits exceeds the cost.

The proposed cost, in terms of hardware and software, education, training and implementation expenses can usually be estimated with reasonable precision. There are likely to be high initial costs, followed by some continuous expenditure for any rental expenses. Benefits will usually not accrue significantly until years two and three, since computer systems of any size usually take between six and twelve months to fully implement.

There are usually several options to consider, not just a go/no-go decision on one expenditure. These options will include alternative ideas of how extensive a system to install at one time, and whether to pay the cost of external skills in order to get the benefits earlier. A common technique is to calculate the maximum figure for benefits, and then to define the project as widely as possible up to the limit of expenditure which can be supported.

One useful perspective on cost justification comes from considering how Henry Ford 1 may have justified the introduction of the moving assembly line in the 1920s. This question also puts into context that, although CIM is relatively new, major innovations in manufacturing industry are 'business as usual'. The question ('How was the moving assembly line cost-justified?') may be worth a few moments thought before reading the next paragraph.

*   *   *

This is actually a trick question. On a totally apocryphal basis, the answer is that Henry Ford sent for his finance people and told them how much money he wanted. He almost certainly did not fill in a form, and then sit back and wait for a decision to be given to him! The corollary of this answer is that when the finance people explained what his plans would do to the financial position of the company, Ford was able to discuss liquidity ratios as easily as direct/indirect ratios, and gearing as easily as line speeds. This answer, let it be repeated, is apocryphal but it hopefully emphasizes that engineers should think the project through in financial terms, and partake positively in business case discussions.

There is a basic questions as to how much of the cost of a flexible CIM installation is to be charged to the products for which it is initially used, and how much of the cost is to be charged to the follow-on products (which, by definition, are as yet undefined). There may be a choice between building a flexible (and more expensive) system, which with appropriate design discipline can be used for the follow-on products, or a cheaper limited system for the current product range only. If the more expensive system is installed, then either a judgment must be made that the increased flexibility will be of real value in, say, two years' time or the present product must be overpriced to pay for the extra cost. However, should the technology change too swiftly, the expensive extra flexibility might have no value, and another new system will be required anyway.

It is much easier for finance people, understanding little about the technology and unwilling to release responsibility for major financial decisions to engineers who understand less about balance sheets, to veer on the side of caution. Thus apparently sound, hard-headed decisions can be taken that each investment in new equipment should be paid for by the products for which it is initially installed. The consequences often include unnecessarily high priced products (which can then be blamed on low productivity?).

It is important to remember that, in the real world, actual commitment of most of the cost is made by the designer. Thus whatever detailed calculations may be made for individual components of it, the CIM system must give support to the designer to use existing systems and production equipment to the full and optimize the design for low-cost production (whilst, of course, maintaining high process and product quality).

Payback period

Much emotion has been spent castigating two rather simple methods for assessing the viability of a project from its cash flows. 'Payback' calculates on the simplest basis the time period before the cost of the project is repaid by dividing the cost by the annual benefit. Payback therefore makes some allowance for the risk of the project, but undervalues those projects that continue to show a high level of benefit after the pay-back period (likely to be between one and two years if the project is to be approved). It is significantly affected by lease/buy decisions on major capital equipment that affect the capital cost used in the calculation, but do not fundamentally affect the viability of the project. 'Rate of return' (as distinct from Discounted Cash Flow (DCF) rate of return) does a simple calculation of average annual benefit divided by gross capital cost. Like payback, rate of return gives no credit for longevity.

Thus, payback and rate of return are essentially short-term criteria, and cost justification of CIM should not be approached on this basis alone, since it is a major business strategy (not just a cost reduction exercise). If cost justification is attempted by calculating how many computer terminals can be bought by saving a storeman, then not only is the justification likely to prove inadequate, but the application may well be constructed wrongly. The pressure of needing the money to pay next week's wages is always present, but it is also important to take a longer term view.

Discounted cash flow

In theory, the best calculation basis is discounted cash flow. This takes account of the time span of a CIM investment proposal on a basis similar to the calculation of interest on a bank loan. In a DCF calculation, future costs and benefits are discounted back to a 'present value'. Expenditure in the present year is taken at face value, whilst costs and benefits in future years are reduced at the rate of, say, fifteen or twenty per cent per year. The arithmetic sum of these present values will give the project a positive or negative 'net present value' (NPV), giving a simple go/no-go criterion. The rate of interest used can be increased to allow for the risk of the project (as compared with putting the money in the bank). Future interest rates (and currency exchange rates if applicable) may need to be estimated.

Using schoolboy algebra, the 'internal rate of return' can be calculated, being the interest rate which will give a net present value of zero. This can perhaps then be used (to push the theory to the limits of reasonableness) to rank projects.

DCF calculations are sometimes also to be done net of tax, which can make the calculation more accurate. However, in many CIM-type systems, tax (as distinct from investment grants or other Government payments) reduces both costs and benefits in roughly similar proportions, so does not usually change the result very significantly; though taxes will of course reduce the amount of money available to invest in the future. However, in some circumstances tax can clearly be important. Except for large Government grants, which are usually more applicable to entire factories than to the CIM systems within them, the most important number in the calculation is the project benefit in the first full year of operation.

Allowing for risk

Some projects fail to give the benefits which were forecast for them; computer systems in manufacturing industry are as exposed to this possibility as other investments. (More exposed by some accounts; there are a number of variations of the famous remark that of the three ways to lose money — often summarized as gambling, drinking and manufacturing industry — the last named is the least fun, but the most certain.)

Two ways of allowing for risk were discussed above:

- the value of the benefits claimed for a proposal can be reduced;
- the interest rate used for discounting can be increased.

The former can be applied more flexibly, and the effect on the calculations is clear; identifiable business risks, such as unexpectedly high increases in raw material costs, or aggressive price cutting by competition (which might each affect just one factor in the benefit calculation) can be taken into account. Increasing the discount rate can be seen as an allowance for the general uncertainty of future events, or for all the benefits of the project arriving later than expected (a not-infrequent occurrence, some might say).

Whatever calculation is used to allow for risk, justification of a CIM project is more likely to be achieved by gaining general agreement on the value of the large strategic benefits, than by discussing the minitiae of assessment methods.

Finding the right people

This is certainly a challenge. CIM requires people integration as well as computer integration. A productivity investigation that begins in the expectation that more robots are required might conclude by recommending a new purchasing system, complete with international communication links to suppliers. Finance people who are not engineers need to communicate with engineers who will be quite unashamed of their inability to read a balance sheet or use double-entry book-keeping. Design engineers need to understand manufacturing processes in depth, and manufacturing engineers understand the design process well enough to make a serious contribution. Maintenance men will have to learn even more about electronics, and production control people will need to use complex statistical scheduling algorithms.

If the people do not have sufficient knowledge and flexibility, there is a danger that the robots will be installed because that's what the people involved understand and feel confident with. External expertise can help (as discussed below) but even so a major education effort is rquired.

Company-based education

Of immediate concern is the education and retraining of existing staff. CIM systems are going to be unique for each manufacturing company, albeit based on common architectures for CIM as discussed in detail later in the book. Even that which is common (e.g. software packages, international standards) will be complex, and mature skill and judgment from all the people in a manufacturing company will be required to run a CIM-based manufacturing business in today's competitive environment.

The challenge of CIM education is twofold: there is a wide variety of new technologies to be learned and also extensive understanding of the business is required. CIM can usefully be compared with MRP in this respect. Education in MRP has been widely available since about the mid 1970s and now includes extensive use of video tapes and self-study texts. Many has been the anxious discussion about the complexity of the subject (Will people be able to understand it?) and how jobs will change. Yet MRP is only a component of CIM, and within CIM needs to include interactive analysis of customer orders and integration with capacity planning, each of these new subjects possibly as large as MRP itself. So if MRP was difficult, what price CIM?

Education needs to be continuous. Systems will change and develop as the users become more familiar with them. Advanced features of a system need to be introduced gradually, building on existing knowledge, as users of spreadsheets or word processing packages are likely to appreciate. As the user learns the features of the package, expectations increase way beyond the application as originally planned. Thus reports will be produced from the word processor replete with coloured histograms and pie charts, and the monthly spreadsheet calculation will evolve to weekly time periods, complee with built-in inflation and risk factors. Use of CIM systems will evolve similarly.

A team approach to most projects will become imperative when CIM is integrating the entire manufacturing organization; teamwork skills will also need to be taught. CIM education is discussed further in Chapter 13.

The standard of the national education system is a relevant factor. For example, Japan produces 80 000 graduate engineers a year, Britain 9000; the relative numbers of engineers per head of population do seem to have an influence on national performance in world markets. There are national and international programmes and initiatives in this area, some of which are discussed in Chapter 13.

External expertise

External experts can make a significant contribution to a CIM project. There are a number of sources of this expertise. Consultants are, of course, widely used; they provide the benefit of a detached viewpoint and a knowledge of what is happening in the wider world. The level of service from consultants, and most other outside agencies, is very dependent on the skills and experience of the individuals concerned.

Another source of advice is the academic world. As the range of new technologies expands, the academic has a real advantage in the opportunity to study these new technologies, and is disciplined to understand them properly by teaching them. Another value of academics is their access to government funds for research projects, thus enabling leading edge problems to be explored in partnership with an industrial company. Academics have the same problem as all external advisors that they are in danger of recommending what they know. Also the natural and proper focus of academics upon the 'interesting' problems can lead to a conflict with the industrialist who needs his problem solved immediately, whether it is 'interesting' or not.

Computer manufacturers provide another source of external information that should be of value in planning a CIM strategy. Advice from the marketing people is likely to be based on a knowledge of future products and computer technology that cannot be communicated directly in a competitive market place. Some computer companies also have the advantage of being seriously in the manufacturing business themselves. As with consultants, however, it is the individual who counts and it is important to establish whether the person who is handing out advice on CIM strategy has any real contact with his own manufacturing people.

A new source of strategic advice on CIM is provided by 'systems integrators'. These companies have usually developed from software houses or consultants, and offer an integration service that can be independent of the hardware and software suppliers. Systems integrators have the specialist knowledge to understand what the different suppliers offer. Like everyone else they may have their own biases in terms of the systems with which they are more familiar.

Value Added Remarketeers (VARs) can also be very productive. These companies are clearly associated with the specific products from selected manufacturers in which they provide installation expertise. In the case of a robot, for instance, a VAR could take responsibility for tooling design, programming and installation. Some large manufacturing companies even use VARs to install their own products in their own factories, a telling comment on the shortage of expertise in some of the new technologies.

Hence there are many sources of external expertise on CIM related techniques. A majority of companies, even the biggest, will use a combination of external advisors. However, a sufficient level of technical expertise in-house is essential to help choose and control the external advisors. Major internal education and training programmes are still required.

Personnel policies

It has been clear in this chapter that well-educated and trained people will be needed for the advanced technology CIM company of the future. Each employee will have to be committed to the business, and to have a wide understanding of how it works. The employee will need to be sufficiently confident of his (or her) position in the company to be able to react positively to a state of continual change. Some of the us/them attitudes which still exist in manufacturing industry will need to disappear. Some of the leading Japanese companies have given a lead in this respect.

Personnel policies will have to reflect (or lead?) this trend, and the current practices of the best international corporations will be more widely applied. Phrases such as 'respect for the individual' will be frequently heard, and this respect probably carried to the level of each person being sufficiently important to justify individual salary assessment and career planning. Opinion surveys (with questions such as 'What do you think of top company management?') will be regularly carried out (answered anonymously!) and published.

Quality

High quality will be essential in all aspects of product and process under CIM, and will be a major benefit from its installation. A cheerful 'try it and see' approach is not going to be acceptable either in defining CIM systems or in using them, since designing a system to cater for mistakes immediately increases its costs and complexity. Much of the real justification of automated systems in manufacturing industry, particularly on the shop floor, is the repeatability which can be achieved once the process is in effective operation. However, quality applies in the process of doing business as well as in the accuracy of component manufacture; in some respects it is an attitude rather more than it is the installation of expensive machinery. Standards such as BS 5750 in the UK (or its equivalent, ISO 9000) will be part of quality-assurance programmes.

Implementation

Implementation of the CIM system is the objective of the whole exercise (and will include specific measures of quality); it is covered in Chapter 12. Before the implementation stage is reached, there are many CIM techniques to be understood. One such technique is the process of analysis of manufacturing organizations, an important factor in determining strategies for CIM; it is discussed in the next chapter.

Further reading

ADAIR, J. (1983) *Effective Leadership,* London: Pan Books

COHEN, S. S. and ZYSMAN, J. (1987) *Manufacturing Matters — The Myth of the Post Industrial Economy,* New York: Basic Books

MILLER P. (1988) *Project Cost Databanks,* London: Butterworths

MORITA, A. (1987) *Made in Japan,* Glasgow: Collins

MORTIMER, J. (1985) *Integrated Manufacture,* UK: IFS Publications

OUCHI, W. (1981) *Theory Z,* Reading, Mass.: Addison-Wesley

PARKER, R. C. (1985) *Going for Growth — Technological Innovation in Manufacturing Industries,* Chichester: Wiley

PASCALE, R. T. and ATHOS, A. G. (1981) *The Art of Japanese Management,* New York: Simon and Schuster

PETERS, T. and AUSTINE, N. (1985) *A Passion for Excellence — The Leadership Difference,* London: Collins

SKINNER, W. (1985) *Manufacturing: The Formidable Competitive Weapon,* New York: Wiley

TOFFLER, A. (1985) *The Adaptive Corporation,* Aldershot: Gower

TOWNSEND, R. (1970) *Up the Organisation,* London: Hodder and Stoughton

VOSS, C. (1986) *Managing Advanced Manufacturing Technology,* Proc. of UK Operations Management Assoc. Conf., Coventry, UK: IFS Publications

Chapter 3

Analysis of manufacturing systems

A review of the applicability of CIM techniques and approaches should include an analysis of the current state of the business's manufacturing facilities and operations. The first section of this chapter describes the various classifications that are conventionally used to describe different types of production system, related primarily to production throughput and technology. The characteristics of these production systems are explored with some comparisons of their advantages and disadvantages.

The second section is concerned with the links between the strategic requirements of a business and the characteristics of production systems to meet those requirements. Increasingly, manufacturing companies need to design production systems that are carefully 'tailored' to business needs, and may require features chosen from more than one of the conventional families of production system. Such tailoring usually requires an evolution of common aims and feasible solutions by multi-function teams. Some frameworks are presented to assist this process.

Finally three types of analysis of current operations are introduced which address respectively human activity, manufacturing engineering and information systems.

Classification of production systems

Most factories contain a variety of production systems that have been designed or have evolved over the life of the plant to meet particular needs. Classifications will not usually represent an entire factory but rather characterize various parts of the operation, in order to establish, for example, whether the approach taken for any particular product line is consistent throughout all its manufacturing operations. Classification also provides a 'shorthand' way of describing various families of manufacturing solution. The most common classification is:

- project,
- 'one off' or jobbing,
- batch,
- line of flow,
- continuous.

Project production

Large and complex manufacturing tasks call for a project approach. Examples include large civil engineering projects and the construction of large-scale manufacturing or military facilities. Work tends to be undertaken not within factories but on sites dedicated to the project with resources delivered to the site and facilities assembled for the duration of the task. The most important tools supporting these activities are various network planning techniques including critical path analysis and PERT. Such tools have been well refined and are widely available as standardized computer packages.

One-off jobbing production

Jobbing shops are set up to provide manufacturing capability rather than a prescribed range of products. Usually a wide range of work will be undertaken to meet a variety of specific customer orders. Production equipment is general purpose and flexible, and the staff will be versatile and highly skilled. The jobbing shop will typically also need strong engineering support as it will be involved with the initial production of new designs or complex one off items such as jigs and fixtures. Costs are likely to be high, reflecting high staff costs and the investment in complex machines that may only be partially utilized. Examples of this type of production would include prototypes, specialist spare parts and test equipment.

Batch production

Systems in this category are designed to produce medium-sized batches of the same item or product. Work to be done on the product is divided into separate operation, and each operation is completed on the entire batch before all the parts are moved on to the next operation. *Figure 3.1* shows the difference between the job and batch approaches. Typically general purpose equipment is still used for batch production, but machine utilization is improved as only one set-up is required per batch. It is important to appreciate, however, that single-minded pursuit of high machine utilization may not always be in the best interest of the manufacturing system as a whole.

It will be apparent from *Figure 3.1* that at any given time only part of the batch is being processed while the remainder lies idle. Also no work is done on parts as they are transported from operation to operation. It has been estimated that in many batch production jobs, parts spend less than 5% of their time in the factory being processed. This has the effect of extending order/delivery lead times, raising work in progress costs as well as imposing complex material control problems and providing extensive opportunities for things to 'go wrong'.

It has been estimated that over 70% of all engineering manufacture is carried out on a batch basis; increasing customer demand for product variety seems likely to increase rather than decrease this percentage. Batch manufacturing has therefore become the focus of activity to tackle its

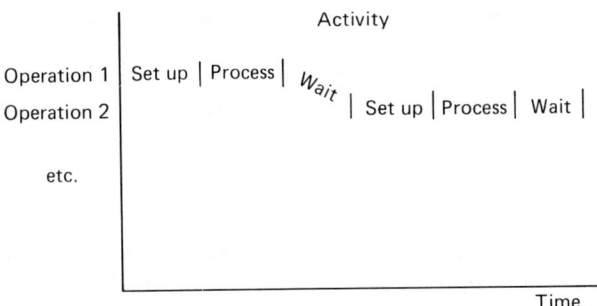

Machine time per component = total set-ups + total process time

Elapsed time = total set-ups + total process + waiting time

(a) Jobbing production

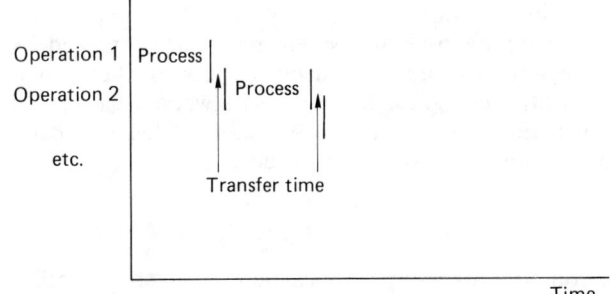

n = batch size

Machine time per component = (set-up time/n) + process time

Elapsed time = total set-ups + (n x process time)

(b) Batch production

Machine time per component = process time + transfer time

Elapsed time = process time + transfer time

(c) Flowline production

Figure 3.1 Throughput implication of major types of production system

inherent problems through more versatile manufacturing technology and novel material control techniques. Developments in cell manufacturing which will be touched on later in this chapter, flexible manufacturing systems and 'just in time' production are all aimed at tackling the problems outlined here.

Line or flow production

When product volumes are high, the markets and designs stable, flow production becomes a viable option. Under this arrangement, parts or assemblies are passed to each successive operation as soon as the preceding operation has been completed. It follows that the throughput time can be drastically reduced and opportunities for mechanization arise with the standardization of products and processes.

A typical example of flow-line manufacture of components is the dedicated automotive transfer line producing engine components. Such systems require a very heavy capital investment and may become obsolescent when product designs change. Increasingly such systems are being build with software-based control systems to allow the maximum possible flexibility while retaining the benefits of the flow line configuration.

Such systems are also vulnerable to disturbances caused by lack of material, breakdown or quality defects. The task of designing a balanced flow line, which requires the operation times for each stage to be equal, is particularly complex in the case of assembly systems that are required to accommodate a varying mix of products.

Continuous production

Continuous production refers to processes, such as oil refining and chemical processing, which are generally distinguished from discrete parts production systems. In order to be technically feasible, continuous flow — sometimes called process — systems have developed sophisticated on-line monitoring and control techniques to deal with variations in such parameters as volume flow, temperature, pressure. The paperwork systems, typical of batch manufacture of discrete parts, are replaced by computer-based systems and the manufacturing processes themselves become automated and controlled, so the differences between discrete part and process manufacture tend to disappear. Arguably, CIM has been established in the process industries for over two decades.

Cellular systems

Batch manufacture of components is typically carried out in shops with a functional layout. Manufacturing processes such as turning, milling and grinding, are grouped together to form sections with a wide capability in each process. The advantage of such a layout is that a wide range of different products with different process requirements can be supported simply by changing the routing through the plant. The major disadvantages include the sometimes complex and time-consuming materials handling and the difficulty of reconciling different scheduling requirements of different products on the various sections.

Cellular layout of manufacturing facilities is becoming increasingly popular as a way of overcoming some of the inherent problems in conventional batch manufacturing. The approach involves the identification of families of components which have similar characteristics. Early examples of this approach, often under the name of group technology, used complex coding systems to identify parts and allow analysis to reveal common geometric features. More recently, production flow analysis has been used as a means of identifying families. This approach uses a plant's standard process-planning data to identify families of parts with similar routings.

In either case, the desired result is the grouping of production facilities into cells capable of completing all the operations on parts within a specified family. This arrangement has a number of advantages:

- the production control task is greatly simplified as components visit only one cell rather than a variety of processes around the plant,
- materials handling is dramatically reduced and throughput time shortened with corresponding benefits from the reduction of work in progress,
- machine set-up times can be reduced as the tooling and fixturing changes between parts of the same family are generally simpler than changeovers between unrelated parts,
- organizational structures can be developed that give staff who operate cells a considerable degree of autonomy and flexibility within the broader but strictly accountable targets set for each cell. Conflicting schedule priorities, shortages and breakdowns can, it is claimed, also be much more swiftly and appropriately resolved within the cell environment. This can lead to manufacturing systems which are very responsive to changes in the mix of demand.

Flexible manufacturing systems

Flexibility in manufacturing systems can take many forms, the most important being the ability to respond to different *volumes* and to different *types* of product. There are no generally accepted measures of flexibility and it is therefore particularly important that system requirements are carefully specified so that the nature and cost of the desired flexibility can be fully evaluated.

The term *flexible manufacturing system* (FMS) has come to mean a group of automated machines capable of processing a variety of products through different process routes under full computer control. Clearly the analysis needed to identify parts families is very similar to the routines for cellular manufacture outlined above. It is, of course, possible to achieve flexibility without automation.

Implications of production system choice

In any manufacturing business, investment in the manufacturing system is substantial; not only in hardware but in layout, training and increasingly in

Table 3.1 Product/service implications of process choice (from Hill, 1983, by permission)

Product/service aspects	*Typical characteristics of process choices*				
	Project	*Jobbing, unit, one-off*	*Batch*	*Line*	*Continuous process*
Product/service range	High diversity	————————————————→			Standard
Customer order size	One-off	————————————————→			Large
Volume of operations	One-off	————————————————→			High
Degree of product change accommodated	High	————————————————→			Nil
Make-to-order*	Yes	Yes	Some	No	No
Make-to-stock*	No	No	Some	Yes	Yes
Ability of operations to cope with new developments	High	————————————————→			None
Orientation of innovation —process or product	Product	————————————————→			Process
Performance criterion: dominant	Delivery/quality	————————————————→			Price
least important	Price	————————————————→			Product customization
What does the organization sell?	Capability	————————————————→			Products

* A make-to-order product/service is one made to a customer's requirement. It will not normally be reordered or, if it is , the time delay between one order and the next will be long. Make-to-stock products are standard items. However, where a standard item is only made or provided on receipt of a customer's order, this is still classified as a make-to-stock item. This is because it is the standard or nonstandard nature of the product/service which is being described by the phrases make-to-order or make-to-stock.

Table 3.2 Investment and cost implications of process choice (from Hill, 1983, by permission)

Investment and cost		*Typical characteristics of process choices*				
		Project	*Jobbing, unit, one-off*	*Batch*	*Line*	*Continuous process*
Amount of capital investment		Low/high ——→ Low		————————————→		High
Economies of scale		Few	None———————————→			High
Level of inventory	components and raw materials	As required ——————————————————→				Planned with safety stocks
	work-in-progress	High	High	Very high	Low	Low
	finished goods	Low ————————————————————→				High
Operations	material	Low/high ——→ Low	————————————→			High
	labour	High ———————————————————————→				Low
Opportunity to decrease operation's cost		Low ———————————————————————→				High

Table 3.3 Production/operations implications of process choices (from Hill, 1983, by permission)

Production/operations implications		Project	Jobbing, unit, one-off	Batch	Line	Continuous process
Flexibility of the operations process		Flexible ———————————————————————→				Inflexible
Set-ups—number and expense		Variable	Many, but usually inexpensive —————————			Few and expensive
Capacity scale		Small ———————————————————————→				Large
Changes in capacity		Incremental ——————————————————→				New facility
Nature of the process technology		Universal ———————————————→		General purpose ——————→		Dedicated
Dominant utilization—labour or plant		Labour ———————————————————————→				Plant
Knowledge of the	operations task	Variable	Known but often not well-defined ——————→			Well-defined
	materials requirement	Known at tendering stage	Some uncertainty ———————————→			Known
Materials handling provision		Variable	Low ——————————————————→			High
Internal span of process		Wide ———————————————————————→				Narrow
Control of operations		Complex	Complex, depending on the product/service	Very complex	———————	Straight-forward
Control of quality		Informal-spot checks ——————————————————————————→				Designed into the process
Process times		Long ———————————————————————→				Short
Capacity	definition	Variable ———————————————————————→				Established
	control	Difficult ———————————————————————→				Easy
Productivity control		Difficult ———————————————————————→				Easy
Bottlenecks	nature	More frequent and random ——————————————————————————→				Very few and known
	position in the process	Moveable ———————————————————————→				Fixed
Impact of breakdowns		Variable, depending upon the importance of the operation	Little ——————————————————→			Enormous

Typical characteristics of process choices

the supporting computer systems. It is this important to examine in detail the implication of production system choice; *Tables 3.1–3.3* outline typical relationships between process choice and product, service, investment and operations. These tables should be seen as a vehicle for exploring the implications of various choices ideally within a multidisciplinary team as outlined on page 44.

Linking manufacturing strategy and systems

Traditional approaches have tended to place decisions about manufacturing at the end of the cycle of strategy development. Production departments were expected to be able to respond to changing market and product strategies frequently across a wide range of models and variants. The conflicting demands placed upon the production engineer inevitably lead to organizational complexity and frequently poor utilization of resources. Increasingly companies are appreciating the need to develop manufacturing strategies in conjunction with other strategies of the business. Indeed the ability to understand and exploit manufacturing is a means of gaining competitive advantage, either through low costs, design quality, excellent service or aspects of all of these.

The process of strategy formulation is iterative and perhaps closer to a craft than a science. Certainly, there are no simple equations that enable strategy to be developed from a given set of inputs. Successful manufacturing companies take the strategy formulation process very seriously. Three frameworks which might be used as part of the strategy development activity are as follows.

Industry characteristics

In the early states of the analysis of a company's strategy, it is helpful to generate some 'rough cut' views of the business characteristics and requirements. There are, of course, many dimensions of business that could be selected, but a matrix of complexity against uncertainty has been found to give useful results. *Figures 3.2* and *3.3* give examples of this matrix and the types of industries that fall within each category. The implications of each market can then be assessed in terms of product, process, organization and information, and a picture can be built up of the performance that will be necessary from the manufacturing activity.

One of the most important strategic concepts to emerge from recent years has been that of 'focus'. This involves bringing together product, process, organization and information systems for each market to be served. Substantial benefits have been observed to follow by avoiding conflicting aims and enabling staff to develop particular expertise in their market. Plants within plants can be created, and although some of the economies of scale of a large integrated manufacturing facility may be lost, benefits in terms of flexibility and appropriateness have been found. There

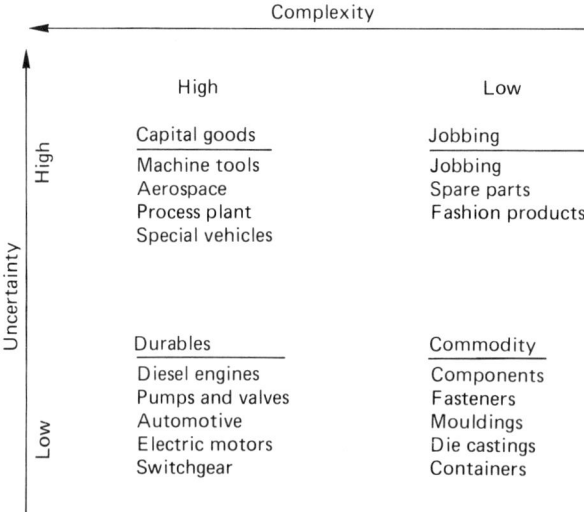

Figure 3.2 The range of market needs (from Puttick, 1986)

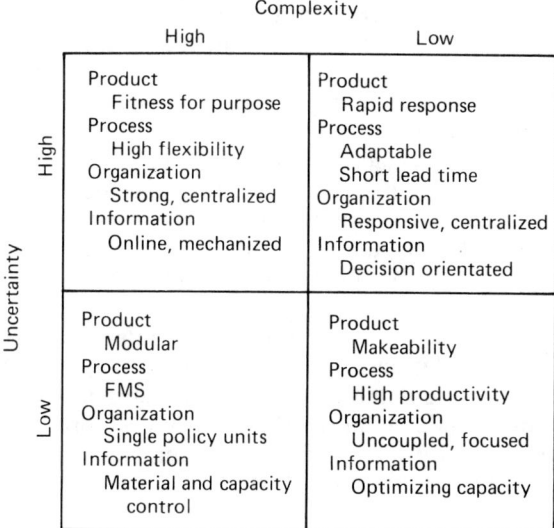

Figure 3.3 Summary of key aspects (from Puttick, 1986)

is a link between the strategic benefits to be gained by focus and the technological benefits emerging from cell manufacturing approaches.

Product life cycles

Any examination of a company's activities should involve a review of the major product lines. The concept of product life cycle is useful here to develop an understanding of the importance of various lines, particularly their contribution to profits, but also their importance in developing future manufacturing requirements. *Figure 3.4* shows a typical growth and decline pattern of a product over its life.

During the introductory phase, products will be needed in relatively small volumes, and will probably be subject to design changes as market reactions are incorporated in new models. The risks of failure through poor

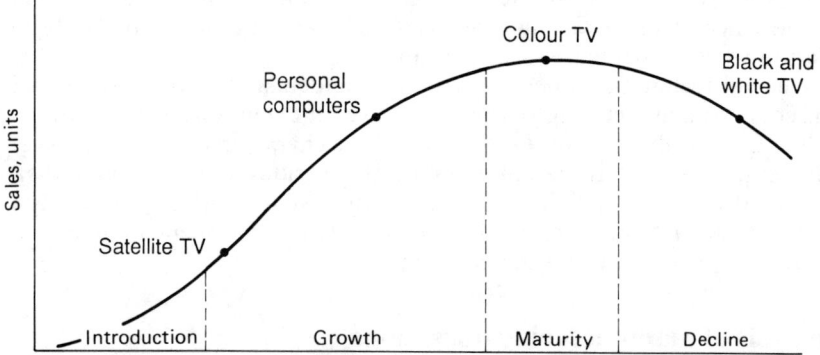

Figure 3.4 Typical S-curve of the introduction, growth, maturity, and decline in the life cycle of products or services

acceptance in the market or unexpected competition may be considerable but the rewards lie in higher prices which can be charged. At this stage generalized manufacturing equipment might be applied to reduce initial start up costs and avoid the loss which might follow an unsuccessful product produced on dedicated machinery.

During the growth period of the product, competitors may be expected to enter the market and higher volume manufacturing facilities with flexibility to accommodate product evolution would be appropriate.

During the maturity phase, with technology and design principles well understood, low cost manufacture is likely to become the major competitive issue.

As product life cycles shrink it becomes increasingly important to understand the structure of life cycles and competitive threats.

Given the different manufacturing implications of various stages of the life cycle it is most important to understand the stage which each product has reached so that future capacity and system characteristics can be planned and routings through the plant modified where necessary.

Product profiling

Product profiling provides a simple graphical means of using the product process relationship characteristics outlined above. The example given in *Table 3.4* shows profiles of a company's manufacturing on two occasions separated by five years. Manufacturing systems that were set up to accommodate high volumes of stable designs have remained in place, whilst the company's real business has moved to low-volume high-variety manufacture. This graphical representation of the key problem areas provides a vehicle for discussion about the link between company strategy and manufacturing systems.

Multidisciplinary teams for manufacturing strategy and systems analysis

Strategy development for marketing and finance have been developed to the extent that some generally-accepted frameworks now exist. Manufacturing strategy however is still in its infancy, partly because manufacturing specialists are not used to defining and representing the performance of their systems in strategic terms. Also manufacturing strategies and performance measures cannot be assessed in absolute terms but only in relation to other strategies of the business.

A useful technique to develop common understanding of markets and consequent manufacturing options is to use an extension of the profiling technique described above. Perceptions of the characteristics of the market can be profiled by representatives of the various functions and then compared with the capabilities of the current system. This helps towards a common understanding of goals but also creates an agenda for action where the most serious mismatches exist.

Analysis of manufacturing operations

The previous section reviewed several of the frameworks which are used in the attempt to link market and product requirements to the manufacturing

Table 3.4 A profile analysis for a company in which the mainstream products have been profiled. The 1983 profile illustrates the dog-leg shape which reflects the inconsistencies between the high-volume batch process and infrastructure and the 1983 market position (from Hill, 1985, by permission)

Some relevant product/market, manufacturing and infrastructure aspects for this company	Typical characteristics of process choice* and the company's 1978 and 1983 product profile		
	Jobbing, unit, one-off	*Batch*	*Line*
Products and markets			
Product range	Wide		Narrow
Customer order size	Small		Large
Frequency of product changes required the market	Many		Few
Frequency of schedule changes required by customers	High		Low
Order-winning criteria	Delivery/design capability		Price
Manufacturing			
Ability to cope with product developments	High		Low
Production volumes	Low		High
Coping with schedule changes	Easy		Difficult
Set-up characteristics	Many, inexpensive		Few expensive
Key manufacturing task	Meet specification/delivery dates		Low cost Production
Infrastructure			
Organization control and style	Decentralized and entrepreneurial		Centralized and bureaucratic

*The process choices open to the company whose profile is represented here did not include a continuous process.

○ ◐ 1978 company position on each of the chosen dimensions and the resulting profile ——————

● ◐ 1983 company position on each of the chosen dimensions and the resulting profile - - - - -

activity. This section looks at some of the ways of analyzing manufacturing systems in detail and three rather different approaches are discussed:

- human activity (sometimes called 'soft') systems analysis,
- manufacturing systems engineering analysis,
- structured analysis.

Human activity systems analysis

Almost all manufacturing systems involve people and the 'soft systems' approach attempts to analyze on the basis of people's perception and knowledge rather than technology. An example is the potentially differing views of three people associated with the same manufacturing activity.

To the shareholder, the factory may be a means of investing money and possibly making profits, to the employee it is a means of earning a living, and to the customer the means of satisfying his requirement for particular goods. Each of these observers would have a subsidiary set of views of the company which would affect their judgement and behaviour towards it. At a more practical level, a key middle manager might see the introduction of CIM as an opportunity to demonstrate his technical prowess and improve the performance of his section, or as a way of undermining and making obsolete his particular skills and knowledge. Such different but credible views demonstrate how incomplete a technological view of systems could be.

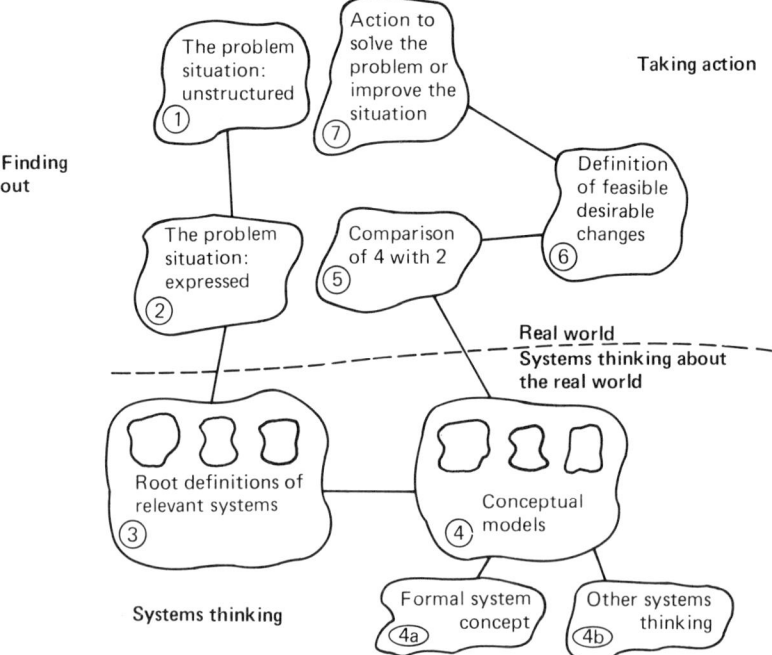

Figure 3.5 The Checkland methodology (Checkland, 1979)

Figure 3.5 shows the soft systems methodology developed by Checkland and Wilson at Lancaster University and applied to a wide range of organizations.

The first stage is concerned with finding out about the situation to be analyzed. The 'rich picture' is developed — literally a drawing showing in cartoon form various activities and personalities within the chosen system, annotated as heavily as possible with their characteristic points of view and any quantitative data which can be obtained.

The second stage of the process involves distilling the essential elements from the picture to develop an understanding of what the chosen system is really trying to achieve together with a statement of how it is going about it at present. The culmination of this phase of activity is the identification of what the methodology calls 'root definitions'. In order to ensure that the definitions are sufficiently well specified the analyst is required to identify within these definitions:

- customers, those people who the system serves,
- actors, those people involved in the system,
- transformation, what changes as a result of the system,
- the point of view which the analyst has taken,
- environment within which the system operates.

Once the definitions have been established, the analyst can consider various ways of fulfilling the objectives implied by the root definitions without the constraints imposed by the existing arrangements. New systems based on the definitions can then be conceived, and compared with the original system revealing inconsistencies, omissions and ambiguities as well as opportunities to be evaluated.

Skill in the application of the methodology needs to be developed, and its value in designing systems to meet clearly specified quantitative targets is unclear. Nevertheless it does provide a way of addressing some of the 'human problems' that are experienced, if not always acknowledged, by practitioners of more conventional 'hard systems' approaches. The discipline of the approach also ensures agreement on basic aims and objectives before expensive system redesigns are initiated.

Manufacturing systems engineering analysis

Manufacturing systems engineering analysis has been developed over recent years, particularly by Lucas Industries as part of a methodology for the fundamental examination of manufacturing operations (*Figure 3.6*).

The manufacturing system engineering approach seeks to tackle a manufacturing system design based on a quantified statement of business objectives. The approach recognizes the importance of the systems aspects of manufacturing, but differs from the human activity approach in that it is specifically goal oriented.

Figure 3.6 gives an overall view of the manufacturing systems engineering approach to design and re-design. The first three sections which constitute the analytical phase of the procedure are:

- business targets,
- market plans,
- engineering analysis.

48

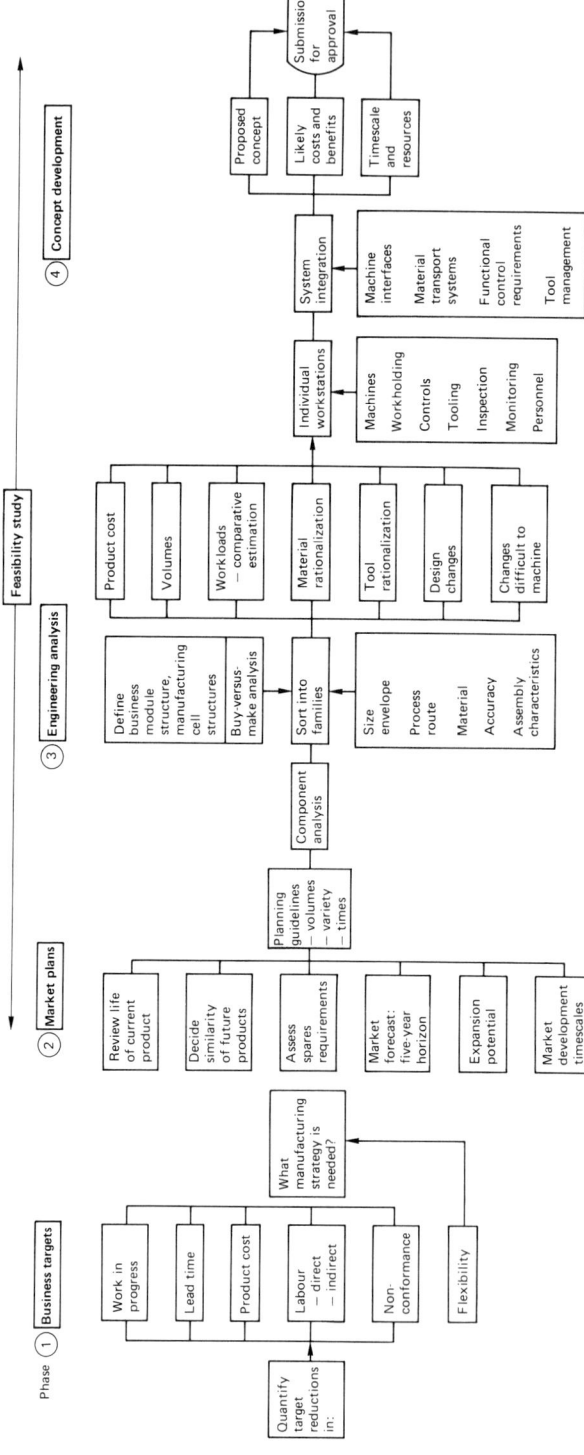

Figure 3.6 Manufacturing redesign study (Open University, 1986, by permission)

The analysis starts from a different perspective to the human activity system analysis, in that required levels of performance are clearly specified at the outset rather than emerging from the analysis of the existing system. Satisfactory overall financial performance is essential, but the operational objectives against which systems can eventually be designed need to be more closely specified. Typical measures would include:

- levels of work in progress,
- lead time or throughput time,
- productivity in terms of people and capital,
- quality levels.

and would be based on a review of the performance of leading competitors.

The second phase of the analysis requires a thorough study of current and potential future markets; in this phase, some of the frameworks, such as those introduced above, are used to provide a view on product life cycles, relevant market sectors, future growth opportunities and potential competitors.

At the heart of the methodology lies thorough and systematic analysis of current manufacturing operations. The first step is to analyze each product line in detail, plotting the process routes for each component and assembly. Process flow diagrams provide a standardized notation for the representation of manufacturing routes and reveal repetitious or redundant operations and opportunities for process improvements. Coupled with this analysis will be a component routing chart showing the movement of material throughout the plant. An important test of activities revealed by the process charting is whether they are value adding. Waiting time or an inspection operation do not add value though they may be necessary consequences of poor system design.

At the factory level it will be necessary to superimpose the process routings for the variety of parts currently being manufactured and this will lead to an overall picture of capacity versus demand for particular facilities. The engineering work to reduce and integrate operations can thus be prioritized and alternative approaches such as sub-contract manufacture explored.

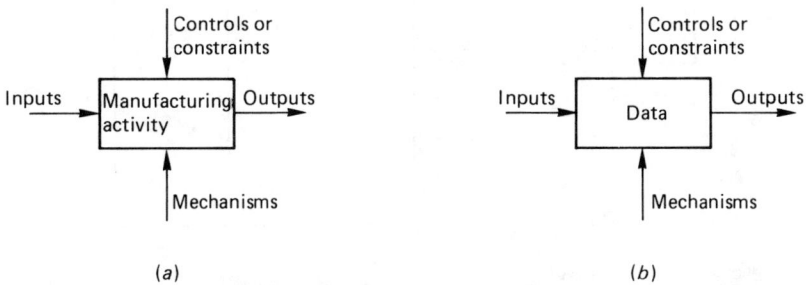

(a)   (b)

Figure 3.7 Basic building blocks of structured models: (a) functional models, (b) information models (Open University, 1986, by permission)

50

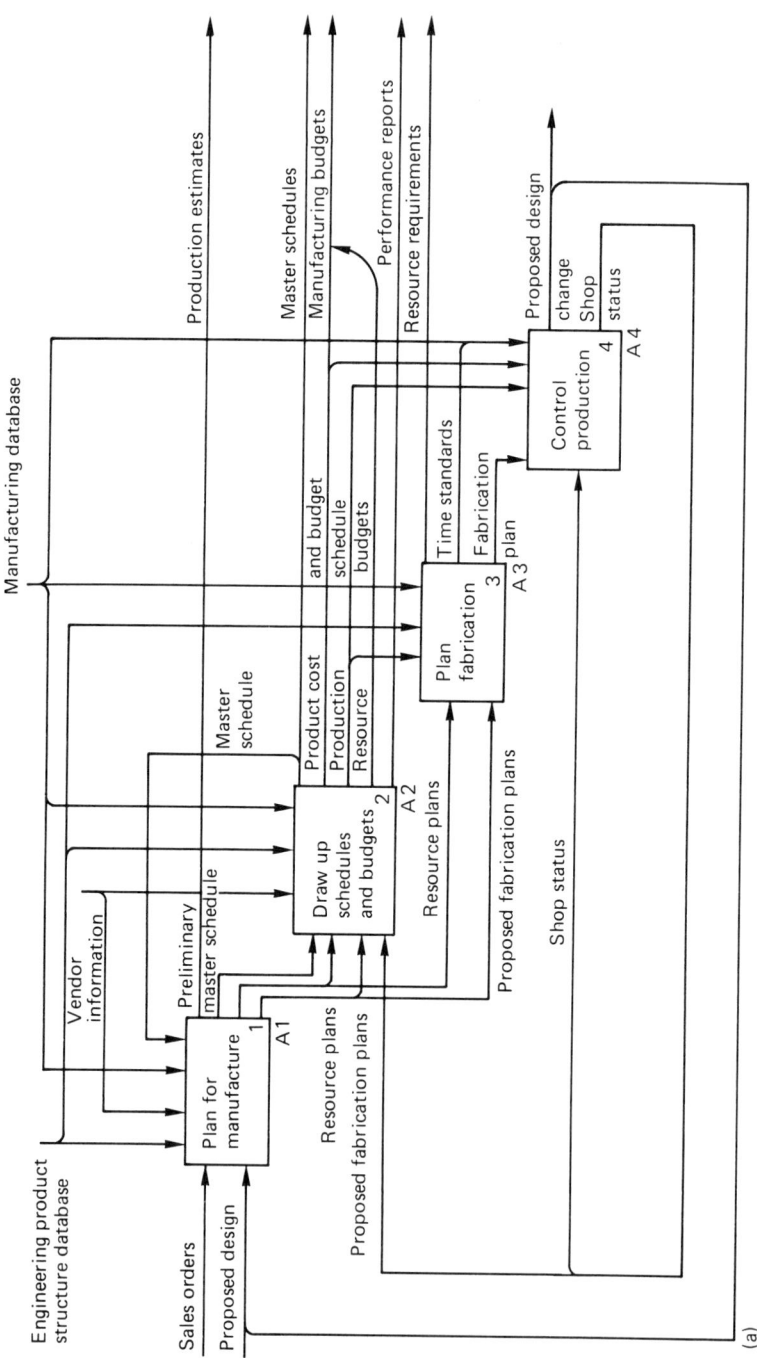

(a)

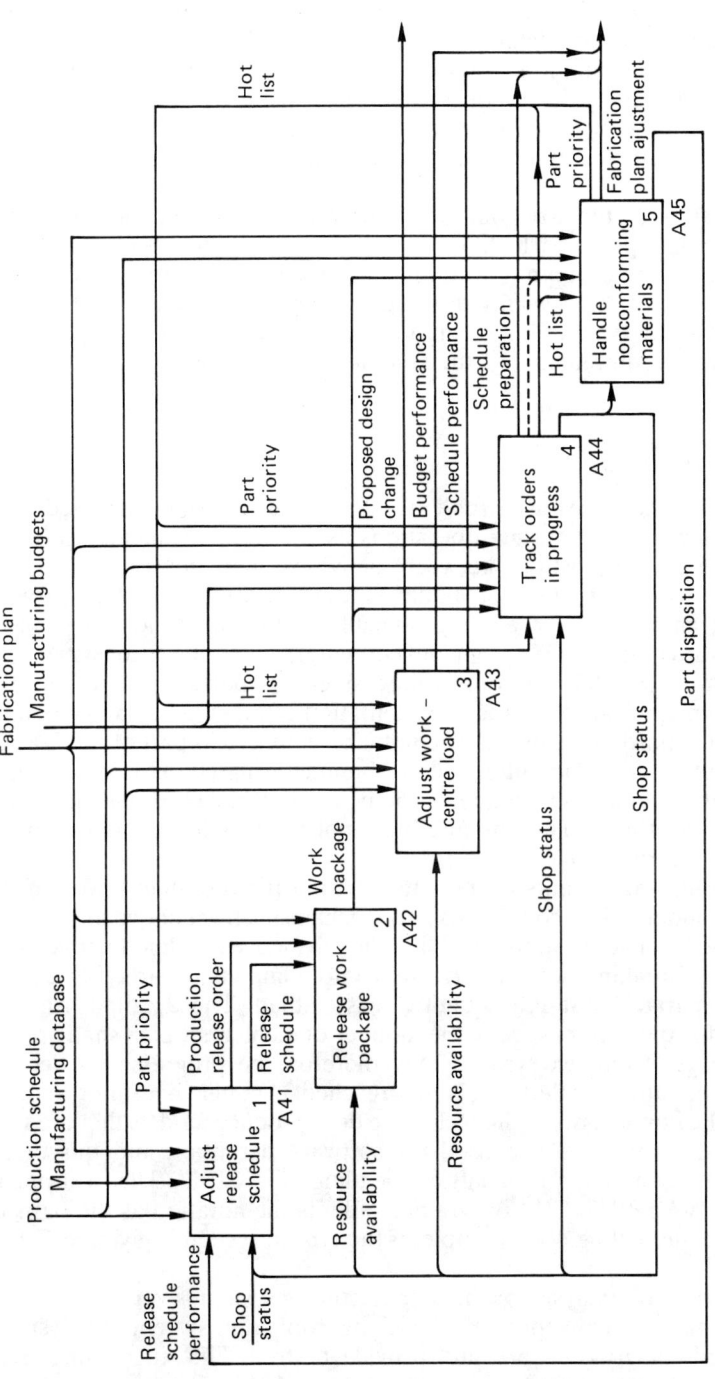

Figure 3.8 The hierarchical arrangement of diagrams: (*a*) a high-level planning and control system, (*b*) decomposition of activity A4 'control production' (Open University, 1986, by permission)

(b)

At the activity level, it is particularly important to analyze the flow of information around the plant; the use of the apparently simple input/output analysis techniques can serve to check the adequacy and continuity of information flows throughout the system from receipt of order to despatch. It is, of course, necessary to select the system boundaries for such activities carefully to ensure that the entire field of concern is mapped. A particular structured approach to this task is given in the next section.

The results of the foregoing analyses provide the foundation for the system design phase, which is likely to involve the re-assessment of processes, the re-grouping of equipment possibly along the cellular lines introduced above and usually thorough re-examination and simplification of the organization and information systems to match the quantified performance requirements of the system.

Structured analysis

One of the most serious problems facing the analyst concerned with representing manufacturing operations is notation. Standard ways of representing relationships between activities and flows of data are important to ensure common understanding of systems and to provide results of analyses which are reproducible and comparable. As projects move towards the system design and computerization phase consistent and unambiguous models become increasingly important. One of the best-known approaches to the representation of manufacturing systems is the IDEF definition language. The notation emerged as part of the US Air Force's Integrated Computer Aided Manufacturing (ICAM) programme and is built around formal models of activities and data (IDEF is the ICAM Definition Method). The building blocks of the activity and data models are shown in *Figure 3.7*.

The inputs and outputs to the left and from the right-hand sides of the boxes are modified by constraints and mechanisms from above and below. This notation in addition to providing the advantages outlined above gives a means of breaking down a large task into component parts that can be tackled separately but subsequently reassembled. The detailed 'rules of conduct' for the system specify the number of elements which shall appear on any page. Complex systems may therefore be progressively decomposed until fully detailed levels are reached. Though it cannot address some of the human issues, the technique does produce models that are very close in structure to those used for software definition and this offers substantial benefits for analyses leading to CIM. The European Community ESPRIT CIM programme adopts the notation as the basis of some of its modelling. An example of the typical system is given in *Figure 3.8*.

This format of analysis for the manufacturing system is very close to the formats that are conventionally used by computer systems analysts to design and specify software and provide a strong link in moving from requirements to computer implementation of CIM. The next stage is to explore in more depth the structure and capabilities of the computers.

Further reading

CHECKLAND, P. (1981) *Systems Thinking, Systems Practice,* New York: Wiley

CROSSLEY, T. R. *Structure Analysis of Manufacturing Systems,* Macclesfield: AMTEC

HILL, T. (1983) *Production/Operations Management,* Englewood Cliffs, NJ: Prentice Hall

HILL, T. (1985) *Manufacturing Strategy,* Basingstoke, UK: Macmillan

OPEN UNIVERSITY (1986) *Structure and Design of Manufacturing Systems* (units for course PT611)

PARNABY J. (ed.) (1986) *Minis and Micros in Engineering and Manufacture,* London and Glasgow: Collins

PUTTICK, J. (1986) *Market Pull,Manufacturing Push,* London: Manufacturing and Logistics Division, PA Consulting Group

Islands of computerization in manufacturing

Chapter 4

Computer system fundamentals

People who decide upon or implement CIM systems need to have some understanding of computer technology. There will be no escaping the need for technical knowledge in the selection, integration and use of computers and terminals. This subject is often referred to generically as Information Technology (IT).

CIM can be said to start when different types of computer systems are linked together; this chapter covers what might be termed pre-CIM systems — single computers (albeit with many terminals) as islands of computerization. Selected aspects of the subject will be covered, since it is (fortunately) not necessary to understand all the internals of computers in order to study CIM. Chapter 7 builds up the architecture for linking multiple computers into CIM systems.

Trends in information technology

It is useful to understand something of the development of computing over the last forty years, in order to provide a perspective for business judgments on likely changes in the future. Different 'generations' of computing are frequently cited, although there are no exact cut-off dates to separate them. Computing began presumably with the abacus, but the first digital electronic computers were developed in the 1940s.

Computers (i.e. the 'hardware') store data and instruction in 'binary' code, which can be represented as strings of bits (BInary digiTs) with values of 1 or 0 (hence 'digital' as opposed to 'analogue' computers, analogue meaning continuously variable). The speed of switching between the 1 and 0 states, and the size of the switch (valve or transistor), is an important difference between the generations of computers.

The first generation in the 1940s used thermionic valves (or vacuum tubes). These were a considerable advance on their electro-mechanical predecessors, achieving speeds of microseconds and allowing programs to be written in assembler languages.

Starting in the mid 1950s, with the second generation, valves were being replaced by transistors (an acronym derived from TRANSfer resISTOR – transistors can be used as either amplifiers or switches). Size and cost were much reduced, and reliability was greatly improved. High-level languages such as FORTRAN (FORmula TRANslation) and COBOL (COmmon Business Oriented Language) were developed. These programs were run in batch mode (i.e. cards were fed into the computer, data files read and updated, and output sent to a printer). Magnetic tapes were mostly used for data storage.

The third generation, in the mid 1960s, introduced integrated circuit (or solid logic) technology, whereby many transistors (perhaps thirty) were packaged on a single chip in one electronic component. The languages of Pascal (not an acronym) and LISP (LISt Processing) were introduced. Cycle times (i.e. speed of the chip) were reduced to nanoseconds (a few thousand-millionths of a second); magnetic disks were used for data storage. Terminals for remote entry of batch jobs were introduced. (Just for the record, when man first landed on the moon in 1969, there were no digital watches and no pocket calculators).

The introduction of the fourth generation is associated with either large scale integration (LSI, perhaps sixteen thousand transistors per chip) in the early seventies, or VLSI (Very LSI) in the mid-seventies. Microprocessors were developed; this technology allowed the development of personal computers (PCs) in the late 1970s.

The fifth generation has no real start point and certainly as yet no end point. It is generally held to encompass the introduction to computing of artificial intelligence and expert systems, and in its full glory requires new developments in hardware.

Parallel processing computers may be an appropriate technology for the fifth generation, possibly to be based on the transputer. For optimum effect, parallel processing requires that problems can be split into multiple parallel tasks; this is clearly feasible for certain functions such as image processing, and is likely to have wider application. Another possibility for fifth-generation computer power is superconductivity, a technology that has recently been boosted by the discovery of new ceramic compounds which function at far higher temperatures than previous superconducting materials (i.e. at $-180°C$ rather than near absolute zero).

Magnetic bubble memory, whereby microscopically small areas of an appropriate crystal (the bubbles) are magnetized, is another possibility. Such memory has no moving parts (a major advantage over conventional disks and tapes) and does not lose the data when the power is switched off (a major advantage over conventional computer memory). Another development is the use of optical disks, further reducing the cost of storing very large amounts of data.

Based on these and other technologies, there seems to be complete confidence that the power of computer hardware will continue to increase for at least the next five to ten years.

It is often noted, sometimes with cynicism, that this extra power is rarely passed in full to the user. However, the user wants more function as well as faster processing, and some of the extra power from new technology is used to provide new facilities for the user, such as high function graphics and the software for large databases.

In development of software, the technology has failed to establish major productivity improvements. Hardware performance has been improving by about 30% a year over the last ten years, whilst programmer productivity has probably only improved by a factor of two in that time (i.e. about 7% per year).

This slow productivity improvement will be a significant limitation in CIM systems, many of which require extensive software. For example, in a CIM system a sales order enquiry might allocate in-stock components temporarily, access suppliers' computers for urgent deliveries of other components and then reschedule machine shop and assembly capacity in detail. It might then apply heuristic procedures to decide whether to take a chance on the components that seem unobtainable and accept the order (heuristics are 'rule of thumb' type procedures). If the order is not to be accepted, all these transactions will have to be 'backed out', which will not be a simple process after a lapse of some time. A great deal of complex software is likely to be needed for such systems.

An important trend in technology covering both hardware and software is the integration of computing and telecommunications (i.e. telephones and associated systems). Telecommunications systems are using technology which is recognizably similar to computers for large scale switching and in order to achieve high traffic rates. Conversely the integration of computers (e.g. between suppliers and customers) makes extensive use of telecommunications networks.

International telecommunications, and communications between different types of computers, both require international standards to define the interfaces. Developments in standards will be a significant factor in progress with CIM.

Whatever the changes in the technology of computing, stability must be preserved for the user. New products need to provide the functions of the old system as well as the new ones, or enable the old system to run on the new computer as an aid to conversion (called 'emulation'). Hence innovations appear as controlled step functions rather than a continuous process of change, but the cost of this stability is high in extra software and hardware.

These themes will be developed throughout the book. A good starting place is the base technology of computer hardware.

Basic computer hardware

The hardware of a computer is a collection of 'boxes' (called a configuration), which includes a central processing unit (CPU), memory, disks, tapes, terminals and printers. Collectively the attachments are called peripherals. The systems come in a wide range of sizes, from microcomputers to large mainframes. It is not necessary to understand in detail how a mainframe computer works; much more relevant to CIM are chips and printed circuit boards.

Chip technology

Buying a computer without understanding something of the chips inside it can perhaps be compared with buying a car without enquiring about the engine size. Small computers are often advertized on the basis of the CPU chip(s) they contain.

A chip is a small piece of a single crystal of semiconductor material containing an integrated circuit (integrated because the electronic components are all on the same chip). The chip is mounted on a substrate, that contains the pins or contacts which connect the chip to the PCB (Printed Circuit Board).

Semiconductors (usually silicon, although other materials including germanium and gallium arsenide are in use) have very low conductivity (compared with copper, for example). What is important is that the conductivity of a semiconductor is very significantly changed by the presence of certain additives (usually called impurities). Impurities are added in minute quantities to the semiconductor. The interface between the pure and the impure (or doped) semiconductor material is a surface through which an electric current will flow or not dependent on a voltage being applied. This is the basis of the transistor, which can operate as a switch; these 'switches' can be placed by the million on a chip roughly the size of a finger nail. Electronic components, i.e. resistors and capacitors, can be produced on the chip by the same technology.

The manufacturing process uses a very pure single crystal of the semiconductor. This crystal will be a cylindrical shape (100 millimetres diameter or larger) and will be sliced into thin wafers. Many chips are usually made on one wafer, which is then cut into the individual chips (though 'wafer scale technology' is emerging, whereby the whole wafer is used as a set of integrated chips). Insulator (an oxide) is deposited on the surface of the semiconductor, and then etched away to give contact points to the diffused regions below. The etching is done by a photographic process called photolithography. Conducting material and further insulators are then deposited on the surface of the semiconductor and etched away to leave circuits connecting the contact points. The circuit connections can be of the order of one micrometre (one millionth of a metre) apart, reducing to perhaps $0.1 \mu m$ in the 1990s.

There are many different types of chips, e.g. bi-polar, CMOS (Complementary Metal Oxide Semiconductor), CCD (Charge Coupled Device), the description of which requires more detail of the technology than is necessary to manage CIM. A gate (as in gate array) is another term for the switches in a chip, and may comprise several transistors. An ASIC (Application Specific Integrated Circuit) is designed and manufactured for one application. The availability of these chips, economically in small quantities, is an important development.

Memory chips
Single chips capable of storing four million bits of memory (Mbits), available in the later 1980s, are likely to develop to 256 Mbits or one billion bits in the late 1990s. Access times of under one nanosecond have been achieved with smaller memory chips.

There are two main types of memory chips. Random access memory (RAM) is used for temporary storage (i.e. only while power is supplied); read-only memory chips (ROM) effectively retain data when power is switched off, but the data cannot be changed. ROM chips can contain control software, e.g. to operate a pocket calculator, and also application software such as a spreadsheet or word processor. Software is a collective word for computer programs.

The economics of ROM chip manufacture are such that for small runs or prototype work they need to be to some extent programmable. PROM (Programmable ROM) chips can only be programmed once, whereas EPROM chips can have the program erased and reprogrammed. The erasing is effected by exposing the chip to an ultra-violet light source. Electrically erasable PROMs are known as E^2PROMs.

Each byte in memory (a byte is a unit of 8 bits) needs to have an address, that is a number usually from zero to the maximum memory size. The length of the number used for memory addressing determines how large the memory can be, and is an important decision in the design of a new computer system. If too long a field is chosen, then an extra work load is introduced in the system during its early life (i.e. an unnecessarily long field has to be manipulated for every memory access). However, a shorter address field length will ultimately restrict the growth of the whole system. The size of memory that can be addressed is also a function of the microprocessor (see below).

Printed circuit boards

The chips, mounted on a carrier, are assembled on to boards or cards; these contain the circuitry, often multi-layer, to connect the various chips together and to interface to other computing systems via 'buses'. As with integrated circuits, a certain amount of faith helps in accepting that the PCB, not much thicker than a postcard and possibly flexible, can have multiple circuit layers, each separated by a layer of insulation. Flexible PCBs can be bent or twisted enabling the final product to be more compact and making it easier to assemble.

Connection of the mounted chip to the PCB can be 'pin-in-hole' (PIH) whereby the pins from the carrier go through holes in the printed circuit board (also PTH, 'plated through hole'), or surface mounted (SMT, Surface Mounted Technology). SMT components, instead of protruding wires, have tabs which are soldered onto pads on the surface of the PCB. High accuracy is required for the SMT assembly process (down to 500 microns between the pads) often making the use of automated machinery or robotics essential. Hand assembly can however be used for pin-in-hole components. The main attraction of SMT is the higher packing density, but since the initial capital cost is also higher, pin-in-hole manufacture is likely to survive for many years.

PCBs and the components mounted on them can be seen by unscrewing the back of any electronic device, such as a hi-fi system, television or computer terminal (after disconnecting the power!) or a pocket calculator. This is an interesting exercise for those not familiar with the technology, if only to see the countries of manufacture of the chips.

Microprocessors

Microprocessors, introduced in the early 1970s (the fourth generation of computing), essentially incorporate the functions of a computer on a small number of chips, or even on a single chip. Using a single chip is not a goal in itself; it is frequently desirable for separate functions, such as input and output, to be controlled by separate chips. Possibly the most famous use of microprocessors is in personal computers.

A simplified diagram of a microprocessor (and in fact many other types of computer) is shown in *Figure 4.1*.

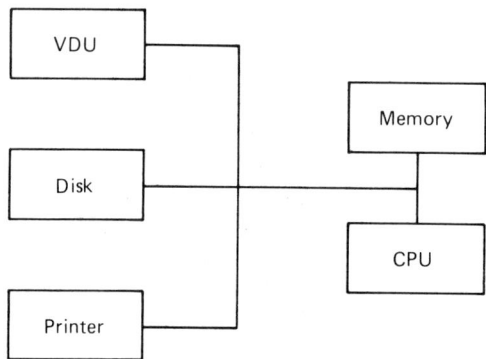

Figure 4.1 Simplified diagram of a microprocessor

The primary control device is the CPU, where the program instructions are executed. The instructions are stored in memory (in a format known as 'machine code') and 'fetched' by the CPU as required. A program consists of a sequence of such instructions, that have usually been written in a high-level language and then converted (compiled or assembled) into machine code.

The power of microprocessors, and hence PCs, can be compared by the width of the data path — although microprocessor chips with the same path width can have substantially different MIPS rates (MIPS, Millions of Instructions Per Second; see page 65). The first microprocessors had four-bit data paths (i.e. effectively four wires in parallel to transmit data). The most widely-used microprocessors have eight-bit data paths; many are embedded in equipment such as cars and domestic machinery. Sixteen-bit microprocessors are logical developments, and in 1985 thirty-two bit devices were announced (containing about a quarter of a million transistors). In some applications there is a limit to the power of microprocessor required, so eight and sixteen bit devices will continue to be used. However many applications in CIM, for example databases and manipulation of CAD data, will use all the power that is available.

Cycle speed is another variable of microprocessor power (measured in MHz, millions of cycles per second), and the faster the clock, the higher is the throughout. However, the clock cannot be speeded up indefinitely, since there are limits in the ability of the material in the chip to change state without losing reliability.

Co-processors can be added to increase the power of microprocessors, particularly in numeric calculations. Co-processor chips are designed for this purpose, and the numbering system often indicates an affinity (e.g. 80386 has an 80387 co-processor chip).

A microcontroller is a special microcomputer adapted to the control of an electrical or mechanical device, and can usually process both digital (numeric values) or analogue (continuously variable) signals. Conversion from digital to analogue and vice-versa (DAC or ADC) can be built in as required.

Microprocessor attachment
Data has to be transferred from one unit of a computer to another; the group of electrical conductors which performs this function is called a 'bus' — literally a set of wires carrying data and control signals

A bus may be internal to a microprocessor, in which case its design is likely to be proprietary to the manufacturer of the microcomputer. However, many buses are external, either for connection to peripherals or to other microprocessors.

Proprietry buses (i.e. owned by a commercial company) arise from a combination of the commercial advantages of owning an architecture and the impracticality of always waiting for the necessarily lengthy consultation processes of international standards to be completed. The technology sometimes moves so quickly that the major companies define their own architecture and announce products. A very public case of this in the context of CIM is the IBM Personal Computer, the interfaces of which have been publicized (and copied) the world over.

Other leading buses include Bitbus (Intel) which is a 'low level' bus designed for high speed serial communication between microcontrollers. Higher level buses include VMEbus (Motorola), Multibus (Intel) and Q-bus (DEC). These buses all provide a published interface which software houses and other manufacturers are able to use.

Interface standards
However, some interfaces are accepted international standards. RS422 Multidrop and the RS232C allow data acquisition and process control equipment to be attached to microprocessors. Common uses will be for transducers and digital counting signals. (The RS Series have been standardized by the EIA, Electronic Industries Association). IEEE-488 (Institute of Electrical and Electronic Engineers) is a well-defined and widely-used standard for connecting programmable devices to computers, originally the General Purpose Interface Bus (GPIB), until adopted as an IEEE standard in 1975.

ASCII (American Standard Code for Information Interchange) is a widely used standard for connecting computer peripherals, based on a seven-bit system. There is a story (maybe apocryphal) that seven bits was simply a compromise between the eight bit bode used by IBM (EBCDIC, Extended Binary Coded Decimal Interchange Code) and a six bit code as an alternative. Whether or not it is over simplified, the story highlights some of the difficulties in agreeing international standards, a topic discussed further in Chapter 7.

Programmable logic controllers

Programmable Logic Controllers (PLC) are microprocessors applied to shop floor control (sometimes just called Programmable Controllers, not to be confused with Personal Computers, even when both are known as PCs!). A typical PLC might count component output whilst monitoring temperature of key processes and controlling an input flow valve. PLCs can operate on a stand-alone basis, or connected with other PLCs or a control computer.

Programming of PLCs is done in a number of ways. A conventional VDU (Visual Display Unit) can be used for input of higher-level languages. English language-like statements can be used with a number of other variants including machine code statements. Ladder Logic Programming permits easy application of the logic of relay operations. The hopefully forthcoming standardization of PLC languages will be not without its benefits.

A controller for a numerically controlled machine tool can consist of a microcomputer and/or PLCs; NC is discussed in Chapter 10.

Personal computers

In hardware terms a personal computer is a microprocessor to which peripherals including memory, a display monitor, keyboard and data storage device are attached, with many other variations and enhanced functions. This brief description hardly does justice to one of the most significant innovations in computing history; however peripherals and software are discussed further below. The unique contribution of the personal computer was to offer serious computing power at a price which millions of users could afford, which generated a deluge of application software to satisfy those users. The current high interest in networking PCs in order to share data and programs is of course consistent with the trend toward CIM.

Mainframe computers

There is a range of terms applied to increasing sizes of computers. Above a micro is a mini, then perhaps a super mini, followed by a mainframe. As hardware, their internal functions are not dissimilar to those of the microprocessor already discussed, with peripheral functions controlled by separate processors, and various other enhancements to improve throughput (e.g. buffer memory to save the time of disk accesses). However, knowing how the internals of a mainframe computer operate does not significantly contribute to understanding how best to use it for CIM.

In manufacturing industry, there is an important distinction to be made as to whether a computer can hold a manufacturing database with Bills of Material, engineering changes, orders etc., available to multiple users in different locations, with full facilities for security and back-up. Such computers would be generally regarded as a mainframes or hosts, although in some configurations a number of minicomputers networked together with the appropriate software can provide the same function.

There are no agreed cut-off points in terms of power or storage capacity for these different types of computer; not really surprising in an environment where last year's 'mainframe' can have the power of next year's 'mini'. However, even though there are no perfect measures, power of a computer system is an important feature.

MIPS, LIPS and FLOPS

Power of computers (of all types) is often measured in MIPS (Millions of Instructions Per Second), which is an approximate measure of the 'number crunching' capability of the CPU. It is approximate because the type of instructions available varies from computer to computer. All computers will have basic instructions, such as for simple arithmetic, but some will have more complex instructions (e.g. double precision division) as well. The total list of available instructions is called the instruction set. Clearly the more powerful the instruction set, the more work is actually done per MIPS (so MIPS counts could be different for two computers delivering the same output). A second limitation with this measure is that actual output of a computer is also dependent on the speed of other hardware (such as the disks), the efficiency of the software and the type of work being executed. However, in the absence of any better suggestion, MIPS is likely to remain as a widely used measure.

LIPs (Logical Inferences Per Second) and FLOPS (Floating point Operations Per Second) are other measures of computer power in, respectively, artificial intelligence (see Chapter 7) and high performance arithmetic (in floating point representation, a number is shown as a base number and an exponent, e.g. $0.42*10^2$). One thousand LIPS are a KLIPS. The S, for second, is not an indication of a plural and cannot meaningfully be omitted, although in practice it quite often is.

RISC architecture

RISC stands for Reduced Instruction Set Computer, in contrast with CISC, denoting a complex instruction set. As discussed above, large instruction sets include complex instructions, the advantage of which is to reduce the cost of compilers and other system software (fewer program statements needed). However reducing the instruction set (i.e. having fewer instructions) reduces the cost of the VLSI hardware. In the technology of the late 1980s, the latter saving is often the more valuable, so RISC hardware is likely to be used increasingly. The CIM user will gain the benefit of the lower cost and/or improved performance, but need not be concerned with details of instruction sets.

Peripherals

These are attached to all sizes of computers, taking advantage of a wide range of standard interfaces. The more powerful a computer, in general the more peripherals it can support, depending on the extent to which the peripherals have their own control hardware and software.

Magnetic tape units

Magnetic tape units have long been the most convenient means of storing data for audit purposes and for back-up (discussed later in this chapter).

Magnetic tapes may also be a convenient means of transferring data between different computers. However, the relatively long time needed to wind through a tape is an obvious limitation in storing data for quick access.

Disk devices

Disks give direct access to large volumes of permanently stored data (hence a frequent acronym DASD, Direct Access Storage Device). Access is via an arm (or arms) which effectively move radially at high speed to locate the track on a disk, and then read or write data stored on magnetic material on the surface of the disk as it rotates. An analogy can be made with the arm of a record player, but there is no surface contact. Several disks will usually be mounted on one spindle (the assembly itself also being sometimes called a disk). Disks can be fixed or removable, indicating whether they can be detached from the computer. Removable disks, although apparently convenient, can add cost because of the need for extra strength and problems of cleanliness. In a typical CIM system, fixed disks will usually be appropriate, since most of the data will be in use most of the time. Fixed disks are often called hard disks.

Data is stored on disks in 'tracks' which are rings on the magnetic surface of the disk. Density of storage is continually increasing. A gigabyte is one thousand million bytes, whilst a terabyte is a thousand gigabytes, an imposing unit of measure to those who remember 7 Mbyte disks in the late 1960s.

Smaller, more portable disks are called floppy disks or diskettes and are used typically with personal computers. These disks can be conveniently removed and stored (even sent through the post). Developments in technology cause periodic changes in the 'standard' size in current use, for which it is well to be prepared.

The technology of compact disks is being used in computing, referred to as CD ROM (Compact Disk Read Only Memory). Music on compact disks is of course stored digitally, so this is a very feasible approach for certain database applications where the data does not change. A single silver CD ROM disk can hold 500 Mbytes of data, sufficient to hold say a year's output of a daily newspaper, i.e. a very large volume of technical data. The technology has evolved to WORM (Write Once Read Many), enabling accumulation of data on the disk (e.g. technical reports as issued or correspondence) and the next generation is developing as eraseable rewriteable optical disks. This is another example of the integration of domestic electronics with commercial computing, holding significant connotations for future control of computing technology.

Video disks, another example of optical disk technology, have proved useful for marketing and educational purposes, although a high volume is needed to justify the initial cost.

Visual display units

There are recognizably different types of VDUs (also called tubes, terminals, monitors, display screens and other terms). A PC can function as a VDU. As a minimum, VDUs consist of a monitor (similar to a TV screen), one or more PCBs and a keyboard, in two or three boxes. There are however many variants, some understanding of which is useful.

The process of entering data through a VDU and receiving a response, maybe on a printer, but more frequently on the same VDU, is an 'interactive' or 'on-line' transaction.

VDU types and compatability
The basic VDU, a 'dumb' terminal, displays data that it has received from a host computer (usually via a 'control unit') or that has been input from a keyboard. It will have a RAM buffer to hold the data while formatting it for the screen, but no means of storing it permanently (e.g. if the power is switched off). As an aid to reducing line costs and improving response time, it may be able to select for transmission the data which has been changed during a transaction.

The measure of the fineness of a VDU display is the number of pixels (an abbreviation of 'picture elements') which are the dots forming the display. The more pixels the better in terms of precision of image (but clearly more pixels means more calculation and therefore a higher powered processor to drive it). A dumb terminal for character display will typically have about 600×500 pixels (allowing for a rectangular screen). In fact, the precision of the pixels, as well as the quantity, affects the sharpness of the image.

A VDU with sufficient power (and the appropriate software) to manipulate user data without recourse to the host computer would be termed 'intelligent'. A typical application would be a CAD model extracted from the central database, updated on the intelligent terminal and then returned, after perhaps several hours work, to the permanent database. Terminals for CAD and graphics work tend to have a higher number of pixels (say 1024×1024), with appropriate extra memory and processing power. At one time, there was a nice sounding balance of a megapixel screen, megabyte memory and one MIPS processor as a high function terminal (probably called a workstation). However, the power and memory which can be economically contained in a VDU is increasing very rapidly (last year's PCB becomes next year's chip, at a much lower price). Chapter 7 describes high function terminals, called workstations, in more detail.

Intelligent terminals, particularly PCs, can usually simulate dumb terminals. Hence, although dumb terminals are likely to be marginally cheaper, the difference may not be great in relation to the cost of a total system; there may be other good reasons within a CIM environment for providing a more flexible terminal. A PC instead of a dumb terminal would give the user not only access to the host systems for which the terminal is primarily intended, but also use of a word processing package or spreadsheet, at no great extra cost.

The major computer manufacturers have defined interfaces for connection of their VDUs that have developed into *de facto* standards. The ASCII standard is a widely used data format for low-function terminals. There are many different types of keyboard, even with one manufacturer, which can be a significant irritant for the user.

VDUs that can substitute one for the other are said to be 'compatible' — not an easy function to define; it is not the same as equivalent or identical. For instance, a terminal with twelve program function keys could be held to be 'compatible' with one having twenty-four function keys; the difference would not matter unless the application required use of more than twelve keys.

Technology of monitors

Most monitors are Cathode Ray Tubes (CRTs). The cathode at the back of the vacuum tube emits a ray of electrons that is deflected by magnetic or electrostatic fields to strike a succession of points in the phosphor with which the front of the tube is coated. This creates the luminescence effect, which needs to be repeated at least fifty times a second to appear steady to the human eye (videotape pictures of VDUs reveal the renewal rate, just as cowboy films show wagon wheels going backwards).

Colour is created by mixing the three primary colours (red, green and blue), with three electron 'guns' and three phosphors providing the three colours. Millions of colours are now theoretically available although only a subset can be distinguished at any one time.

A vector scan VDU has a built-in capability of displaying vectors (i.e. straight lines). This can reduce the programming effort, although the resulting software is specific to that type of VDU. A 'bit mapped image' is now the usual approach, whereby vectors are converted to raster (TV like) screen images in a buffer and then displayed on the screen.

Liquid Crystal Display (LCD) is used where low-power displays are required for portable (i.e. battery driven) computers. This provides adequate readability in most environments. Flat display screens, or panels, also use LCD technology and can be less than one inch thick. This reduces the 'foot-print' of the VDU on a desk (and can therefore be a real cost saving) so is likely to become more widely used. Gas panel displays have had only limited use.

Stereo VDUs provide alternative images for the right and left eye, which are combined using polarized glasses.

It may be too elementary to mention, but just for the record the reason TV does not require the same calculation function as a VDU is that the image is collected by the camera — it is not created by calculating the colour of each pixel. However, TV programmes and films show many examples of the mixing of pictures and computer graphics. The technology of the display tube in a TV set is similar to that in a VDU, which (as with disks) has implications for world manufacture of computer hardware over the next decade or so.

WIMPs and other devices

Keyboards are the most frequent means of entering data to a VDU, and because of their flexibility are likely to remain so. A set of other input devices is known by the conglomerate acronym WIMP. Windows are small display areas on the screen, usually rectangular, which show multiple screens simultaneously. If a list of options is in pictures rather than words they are referred to as Icons. A Mouse is an alternative to a cursor for selecting options from a menu, or for example a feature from a CAD drawing; it is a hand-held device controlling a pair of intersecting guide lines on the screen. Pull-down menus offer a selection of options.

Touch-sensitive screens can be used for selection from a menu, but the lack of precision of the human finger tends to limit their use. A light pen can also select fields from the screen.

Methods for entering data specifically in the manufacturing environment are discussed later in this chapter.

Printers and plotters

Dot-matrix and daisy wheel printers are impact printers (the daisy wheel being a removeable wheel with all the print characters on it). Dot-matrix printers form the print from a rectangular matrix of dots; they are thus more flexible than the daisy wheel printer, can print graphics and are usually faster and cheaper.

The most common non-impact printers use laser technology, followed by ink-jet and thermal printers. Laser printers produce high speed and high quality printing. The power and flexibility of laser printers requires special software, often known as Page Description Languages (PDL). These languages may be either on a host computer or on an 'intelligent' printer.

Special large-size printing may be needed in the manufacturing and distribution environment so that labels can be seen from a distance (e.g. from the top shelf of the stores).

Plotters are used principally for the output of CAD systems, and also in business graphics, including drawing on film for use with overhead projectors. There are two principal types of plotters, pen plotters and electrostatic plotters.

Standardization in printers and plotters can be achieved by flexible interface software, the printer then adapting automatically to whichever protocol is being used. The use of printers in 'in-house publishing' is described in Chapter 11.

Shop floor communication

Communication on the shop floor is clearly an important aspect of CIM. Data needs to be collected and instructions transmitted in such a fashion that the people on the shop floor have adequate control of the technical and business processes, and are also appropriately involved with decision taking.

It is not cost effective or accurate for data such as part number, batch number, operation number and quantity to be typed character by character into a computer terminal. Accurate collection of data requires a degree of automation; this can be achieved by coding the data into a magnetic stripe or bar code, for instance, or by displaying it on a terminal for the user to select.

The types of terminals described below need to be linked to each other, and (presumably) to the host computer system. This will usually be achieved by a shop floor network, to which the terminal may be connected directly or via a controller. Network technology and applications are described in Chapters 7 and 9.

Shop floor terminals
Terminals often need to be 'ruggedized' for shock, vibration and rough handling on the shop floor. Temperature extremes from sub-zero to 50 or 60°C need to be accommodated. Special keypads may be required for usage with heavy gloves. Electromagnetic radiation may be a problem (for cabling as well as the terminals); air filtering is likely to be essential, and also electrical insulation in dusty environments to avoid the risk of explosion. There is a high emphasis on twenty-four hour a day reliability, including battery back-up and capability to withstand power surges. Large size display may be needed on terminals, to be visible across the factory floor. These considerations may require major redesign of conventional terminals; because of their relatively low volume, shop floor terminals will usually be more expensive than the office equivalents.

Barcode readers
Barcode labels can be printed on most printers with ordinary text, and allow data to be easily read with a hand held 'wand' attached to a terminal or by a laser beam (as seen in many supermarkets). However, barcode labels are sensitive to dirt, so are best used in clean manufacturing environments or where the documentation is only intended to have a short life. There are different codes for barcode readers. Other techniques exist for printing data on paper in a machine-readable format, but they generally have the same need for clean conditions and careful handling.

Magnetic stripe
Magnetic stripe technology (as used in many credit cards) is well suited to the shop floor environment since data can be read from a magnetic stripe even if it has been soaked in oil or painted. However, production of magnetic stripe documentation is more expensive than bar code since special printers or documentation are required.

Transducers
Test results, weights, flows, counts, temperature etc. can all be collected directly by computer with the wide variety of low power and accurate transducers now available. Transducers, usually connected to a PLC or microcomputer using a two-wire interface, are discussed in Chapter 10.

Smart cards/intelligent tags
Smart cards and intelligent tags have sufficient memory to hold useful data about jobs progressing through production processes (e.g. routing sheets, tool requirements). The information in a smart card/intelligent tag can be displayed on the shop floor without the need to access a computer, a major advantage in some situations. However, this information will usually need to be duplicated on the central computer for the people taking production control decisions, e.g. about future work load and anticipated completion dates. Duplication of data has the inherent risk that the two versions can (some would say will) get out of step. This may lead to faulty decisions, or at best to the cost of a manual checking exercise to correct the bad data. Provided there is agreement as to which is the master data, use of smart

cards/intelligent tags can be beneficial, if for some reason there is no access to the computer system (e.g. work is off-site, large finished products being stored in the open).

Sound/speech
Computers can now recognize thousands of words, such as for example verbal input of part numbers and stock levels. Computers can also synthesize speech, i.e. the computer could answer back to agree or disagree with the stock level! In spite of the apparent convenience, the business justification and particularly the accuracy of such techniques needs to be established.

A possible application of sound monitoring in the operator-less factory will be to record the general noise level as an indication of whether production is proceeding smoothly (nasty screeching noises or unexpected silences being indications of problems). However, this would be a rather crude method of production monitoring, and contradicts a spirit of 'get it right the first time'.

Vision
The increasing use of vision within assembly and other production processes will be discussed in Chapter 10.

Access security
Physical access security can be controlled with the familiar identity badge readers on locked doors, with computerized passwords and even voice detection available for a higher level of security. Access to computer data (particularly personal data) can be controlled with the same badges, by making entry of the badge into a reader attached to the terminal a part of the transaction.

Telecommunications

The terminals discussed above will need to be connected to a controller and/or to a mainframe computer. The connection is generally described as 'local' if terminal and controller are both on the same site, i.e. the connecting cable does not cross the public highway. Connection between different sites is termed 'remote'. Remote communication involves use of the Public Switched Telephone Network (PSTN) or a Public Data Network (PDN); the organization that runs this is usually a PTT (the national telecommunication authority), either Government owned or tightly regulated.

Initially, transmission of data was on ordinary speech lines, based on analogue technology. Computer data, being in digital (i.e. binary) format had to be converted into analogue format; this was done by modems (MOdulator/DEModulator) at each end of the line. The speed and quality of such lines was set by the requirements for voice transmission; neither needed to be very high (one conversation per line by humans who could often interpret a message even through the noise of static). The effect on computer links was to limit transmission speeds, although computer software enables these lines to be utilized as efficiently as possible, by only transmitting that part of the data that had changed, and by compressing text data.

There are different types of transmission, which are worth understanding in outline, although the details need not concern the CIM student. The problem is to transmit accurately — without losing occasional bits — and at low cost. 'Asynchronous' transmission (also called 'start-stop') sends a checking signal at the beginning and end of each character; the technology is simple, but only low speeds (e.g. 1200 bits per second) can be achieved. 'Synchronous' transmission controls signals with timers at each end of the circuit; the technology is more complex, but higher speeds can be achieved. 'Full duplex' is simultaneous transmission and receiving, 'half duplex' is one or the other at a time (but not both).

Over a period of years, voice networks are being converted to digital operation (ISDN — Integrated Services Digital Network — a long time is required because of the enormous cost). This conversion is transparent to someone making an ordinary telephone call, and is much the cheaper way of handling both voice and data (for instance, modems will no longer be required for the sending of data).

Protocols for public telecommunications are 'recommended' by the CCITT (Consultative Committee for International Telegraphy and Telephony). For example, the CCITT V24 recommendation is used internationally as a terminal interface — more precisely for circuits between a DTE (Data Terminal Equipment) and DCE (Data Circuit-terminating Equipment). This interface is also known as RS232. Other V and also X recommendations relate to telecommunications protocols.

Further protocols are needed if multiple terminals are to share a telephone line, since the switching of data on the line between different terminals is then required. One such technique is called 'packet switching' whereby 'packets' of data are sent down the line, as in *Figure 4.2*. This shows a simplified representation of data packets, with computers C1, C2 and C3 sending data along the same line to terminals T1, T2 and T3.

Multiplexers (a generic computer term for simultaneous transmission of signals on one communications channel) separate and rejoin the packets so that each transmission is cohesive.

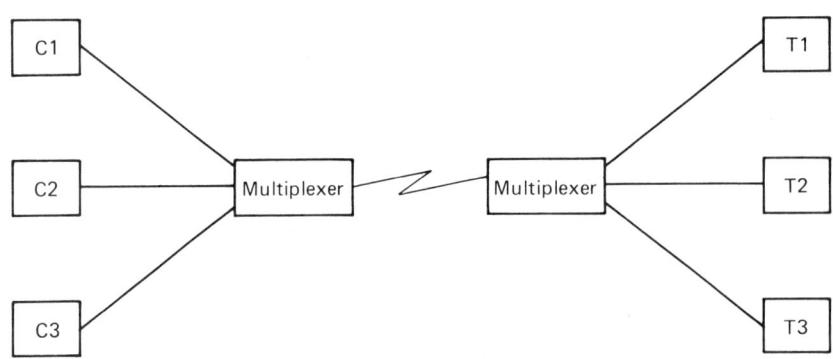

Figure 4.2 Packet switching

The X25 protocol

The principle CCITT recommendation for packet switching is called X25. It defines a three level protocol, consisting of a physical level, a frame level and a packet level. X25 enables 'packet terminals' (i.e. a specific type of terminal) to communicate via 'packet switched networks' (which may be public or private), and also allows conventional (character) terminals to use such networks with an interface box called a PAD (Packet Assembler/Disassembler). The X25 recommendation was first issued in 1976, and modified in 1980 and 1984. It is accepted internationally.

Software

Software is a generic word for computer programs of all kinds. The important content of what the user sees is controlled by the 'application' software; this is however only a part of the total software required. Below the application software are different types of 'system' software, including operating systems.

Operating systems

An operating system is usually supplied by the manufacturer of a computer; it provides functions such as scheduling jobs through the computer, managing the use of memory and controlling disk access. This is a non-trivial scheduling task; execution of machine code is measured in microseconds (or less), accessing data from disk in milliseconds and input from a keyboard in seconds. These different timescales require careful management (as can be understood by increasing them in approximate proportion to more familiar time units of minutes, days and years).

The user does not need to be aware of exactly where the operating system ends and other software begins (e.g. language compilers or a database manager). *Figure 4.3* shows the relationship of the different software components in computer memory.

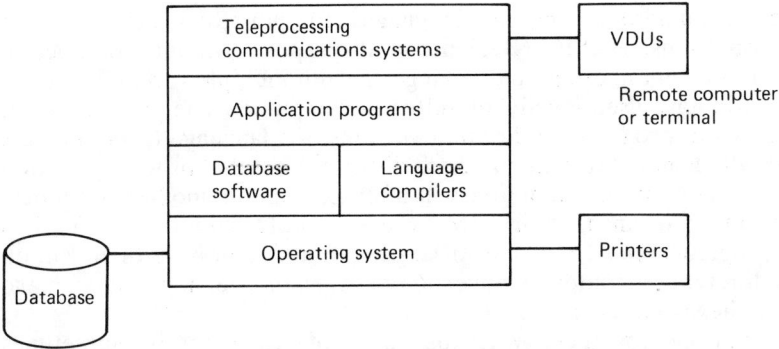

Figure 4.3 Software components in memory shown diagrammatically

The simpler operating systems, such as the disk operating system on small PCs, can only run (or 'execute') one program or 'task' at a time. Large operating systems can simultaneously handle multiple user jobs, network control and database management, for instance, including both 'batch' and 'on-line' programs. As discussed above, on-line programs (also called interactive) are those that respond, usually within a few seconds, to input from a terminal (the process being called a transaction); batch programs take longer to run. (The distinction between batch and on-line used to be much clearer in that a batch program was started by feeding punched cards into a card reader. However there are not many punched card readers about now, since batch programs are usually initiated from terminals.)

Some operating systems provide an important function known as 'virtual computing', whereby each user can appear to have disposal of an entire computer, even though there are in fact many users on the one system. This effect requires additional complexity in the operating system, and part of the CPU power will be used on this extra overhead. Virtual memory, for example, is managed by the operating system copying blocks of RAM memory (say 4K bytes at a time) onto disk and back into memory as required. Thus when the program currently being executed requires more memory, the contents of the block of memory that has been least recently used (or selected by some other algorithm) are sent to disk by the operating system. When next needed (probably within a few seconds or so) that block of memory can be reloaded from disk to real memory in ten to twenty milliseconds. Since memory (in this context called 'real' memory) is measured in megabytes (10^6 bytes), and disk space in gigabytes (10^9 bytes), many virtual memories can be handled if required.

'Utilities' are software for frequent house-keeping functions, such as copying files, sorting records in sequence, maintaining program libraries etc. They are usually provided as part of the operating system.

Although operating systems can be both complex and interesting, the choice of operating system (and there are many of them) should be determined by what the user wants, and not vice versa. Of more immediate interest in a CIM context are computer languages and databases.

Computer languages

There are numerous computer languages, for most of which there are many variations. BASIC (Beginner's All-Purpose Symbolic Instruction Code) is a common example of a language with multiple variations; many people will have used BASIC on different computers. The problem is not the writing of programs in different versions of a language (programmers will usually handle this without much difficulty) but that programs written in one version on one computer will often not run on another computer. This is an important issue in CIM, where it would often be beneficial to write programs once (in whatever language), and then have them run on many different computers. Conversion between one version of a language and another is rarely simple.

The reasons for these variations are not, as might be sometimes supposed, either sheer awkwardness or commercial defensiveness on the part of the suppliers. The problem is the need to get the best out of each particular computer/language combination, and of not being able to put all the features of a language on a small computer.

The option is always open to programmers to use only the features of a language that are likely to be standard in all versions of it. This in effect means using only a subset of the language, and not taking advantage of features that run only on particular computers. The possible future benefits of program transferability have to be balanced against the short-term disadvantages of longer, and hence more expensive, programs.

High-level languages
Three high-level languages were mentioned at the beginning of this chapter. BASIC is widely used on home computers but is generally held to lack some of the facilities necessary in a professional programming language, particularly in database handling and suitability for structured programming (see below). COBOL is the most commonly-used language, but is weak in handling mathematical calculations (reasonably enough as it is intended as a commercial language). A FORTRAN version called FORTRAN 77 is the most-commonly used language for scientific and engineering work. It appears in some respects similar to BASIC, and once learned tends to be not forgotten.

PL/I is another high-level language. It combines some of the features of both COBOL and FORTRAN. Pascal (not an acronym) is much used for the teaching of programming. It encourages a structured approach to programming, as do two further languages, Modula-2 and Ada. Ada has strong commitment from Governments as a 'real time' language for defence systems ('real time' being the continuous immediate response needed, for instance, to control an aeroplane automatically). Apart from this official backing, Ada has many technical merits, and is likely to rapidly grow in importance.

APL (A Programming Language) is a different type of language. It was invented as a mathematical notation and can be recognized by its use of symbols such as arrows and boxes, with the occasional Greek letter looking gratifyingly familiar. APL is interpreted rather than compiled (see below). Regular use is necessary to keep the symbols in mind, and a special keyboard is required. However, it is concise and extremely powerful for mathematical and scientific work.

The language C has potentially great relevance to CIM because (like Ada) it is intended to be highly portable, that is to say to run on many different types of computer. C is used both for application programming and, because of its speed, for writing operating systems. C is the language used for the operating system UNIX which will be discussed in Chapter 7 as a possible component of a CIM architecture.

Quite different languages are those used for programming numerically controlled machine tools and for robots. APT (Automatically Programmed Tools) is the leading high-level numerical control language (although many numerical control programs are produced direct from CAD systems). In spite of a number of attempts, there is no widely accepted candidate for a standard robot language.

LISP (LISt Processing) and PROLOG (PROgramming LOGic) are

languages used in AI (Artificial Intelligence) and expert systems, and will be described in Chapter 7.

Compilers, interpreters and assemblers
Compilers translate high-level languages, such as COBOL or FORTRAN, into machine code. The program statements that are input to a compiler are called source code, and the output is object code. Object code can be 'executed' immediately (i.e. the program can be run) or it can be stored on disk in a computer 'library' for later execution. Keeping parallel copies of source and object code is a non-trivial task for the information services department (and it is not unknown for the source code of infrequently used programs to be misplaced, making it difficult to update them). Recreating the source code from object code is not easy, so computer packages are sometimes distributed in object code format in order to protect copyright.

Interpreters take source code and execute it line by line. Such programs usually run less efficiently than compiled programs, but the technique has great value in interactive and testing environments, because errors can be immediately found and corrected.

'Assembler' is a low-level of programming that requires more detail in the program than does a high-level language. An assembler language allows control of each byte in memory by its address (for instance an instruction such as MVC — for 'move characters' — will move the contents of an address in memory to another address). An assembler language will, like a high-level language, accept 'symbolic' names, e.g. a field can be called 'STOCK', which makes it much easier to refer to than by its address. An assembler program is translated, by an assembler, into 'machine code', which has no symbolic names, just strings of bits representing addresses, field contents and instructions. Machine-code instructions, the lowest level of program, are usually printed in 'hexadecimal', i.e. a number to the base of sixteen ('hex' for short). Hex consists of the ten numbers and the first six letters of the alphabet (for example, B in hex is equivalent to eleven — binary 1011), hence two hex characters fit into one byte. Hex is a valuable medium for some types of program debugging (a 'bug' is a fault in a computer program), but is not used for writing programs.

Assembler languages and machine code naturally give much more potential for error so tend to be used only by professional programmers in situations where high execution speed is essential, typically in system and communications software. For application programming, increased use is being made of 'application generators', or 4GLs (Fourth General Languages).

Application generators/4GLs
As discussed at the beginning of this chapter, software productivity has failed to match improvements in hardware. The queue for new computer systems in manufacturing companies (the software backlog) can often be two years long, this being the limit of time for which it is meaningful to maintain a queue, rather than the limit of business needs for new systems. In most companies, this backlog is as old as the computer department. Application generators, or 4GLs (also called ADTs, Application Development Tools), may provide part of the answer.

There are many such products; the term 4GL has become a generic description of a set of methodologies rather a closely-definable product. A 4GL will usually include an on-line interface that allows formatting of displays and reports.Thus the user might tell the 4GL programmer, as they both sit at the screen, that the stock value field should be in red if it is below re-order point, and could it be a couple of columns to the left, allowing room to display the 'make or buy' field. This display could be prepared quite quickly. The user may be able to make minor modifications without reference to the programmer. Serious expertize in the 4GL would have been required for the initial decision that it was suitable for this particular application — a company may need to maintain skills in more than one 4GL.

4GLs can provide major productivity increases; factors of five and above have been quoted for ideal applications. However, there can be a cost in run-time efficiency, which may limit their use in systems that are heavily used. The relatively simple logic of which many 4GLs are capable is unlikely to be much assistance to the more complex aspects of CIM, but they may well have significant value in simple transactions such as stock movements or for report formatting.

Systems analysis and structured programming
Whatever language or other software is used, there is a need for someone to look at the business problem and to determine how (and whether) a computer solution can best be applied. The conventional method of developing computer programs is for a systems analyst to write a 'specification' of the program. This is a description of the input, logic and output of each program (the systems analyst's job includes breaking the total problem into separate programs). A programmer then writes and tests the programs from this specification. Systems analysts, whose starting point is the analysis of a business problem, may well not be good programmers; many good programmers are more interested in well-written and correct code than in the overall business problem they are helping to solve. The check and balance of these two different approaches can work quite well.

With very large suites of computer programs, such as in CIM and other major software, it is particularly necessary to ensure that the programs are well 'structured'. This means that it must be clear how the functions of the many different programs relate to each other, logic must not be duplicated (or omitted), program maintenance must be easy and it should be possible to test programs separately.

One technique used to ensure good systems analysis and programming work is a 'structured walk-through', whereby a team of analysts and programmers together review the logic of a set of programs to look for errors and inconsistencies before the programs are written.

An example of a program discipline within structured programming is to avoid use of the 'GO TO' statement. This statement has the disadvantage that when a program error is found (and identified by the line number in which it occurred), it can be difficult to determine what the program was doing at the time of the error. In hundreds of lines of code there may be many GO TO statements that could have sent control to the line with the error. It is often better to write the program in modules that perform specific functions. These modules, or subroutines, can be 'called' when required. The flow of logic of the program will then be much clearer (in the ultimate, a sequence of 'call' statements). Some languages by their nature encourage this sort of structure.

For complex software a mathematical approach to program specification can be used. Z (quite different from the languages so far described, and not to be confused with C) is an example of a language for formal analysis of the system function and program structure.

Database

The requirement to hold, and eventually access, large amounts of data has led to extensive development of computerized databases. For permanent storage, the data is held on disk, where any record can be read within say 10 to 20 milliseconds (it takes a lot longer than this to format the data and send it to a terminal).

A database is a collection of interrelated data, stored so as to be efficiently available for creation, update and deletion, by as many remote and local users as the applications require. Four views of the data can be considered:

* how the data is actually held on the disk, the 'physical' organization,
* the 'logical' organization of the data (the logical data base description is known as a 'schema'),
* the application programmers view of the data,
* the user's view of the data (usually on a VDU).

There are precise definitions of these terms, and much other terminology, provided by CODASYL (Conference on Data Description Languages) and/or the software suppliers. Two main types of databases will be considered in outline in this chapter: hierarchical and relational. The objective will be to explain the subject to a useful level, without too much detail, and to give some perception of its historical development in manufacturing industry; this will assist future decision making.

Hierarchical databases
By the mid 1960s, some leading manufacturing companies were using the forerunners of hierarchical databases, often known as Bill of Material Processors (BOMP). A 'record', describing for instance a sub-assembly, was located on disk by its address, the address being a number held in an index (or sometimes calculated from the sub-assembly number by a randomizing algorithm). The record consisted of a set of 'fields', which contained either user data (such as description, cost, quantity) or the addresses of related records (e.g. a component of the sub-assembly). Hence it was possible to follow a chain of addresses (which need not necessarily be all on the same disk) and extract all of the components of a sub-assembly.

The technology of disks was such (and still is) that, in this type of database application, it is generally more efficient to store fixed length records than to have fewer records of varying lengths. Hence, the record

for a washer will be the same length (in principle) as the record for the
aeroplane into which the washer is ultimately assembled, even though the
aeroplane has thousands of components. The difference between the
records is that the field for 'address of first component' is empty for the
washer (being purchased, it has no components) but for the aeroplane this
field gives the address of the record for a subassembly (e.g. a wing). The
record for the wing points at other subassemblies, thus arriving ultimately
at the washer.

There was considerable discussion (and doubt) at the time about the
business and strategic benefits of BOMP systems, a discussion that is hard
to sympathize with now, when modern versions of BOMP systems can run
on some home computers. Compared with the alternative of storing the
data on magnetic tape, the value of BOMP systems was the speed with
which the data could be accessed on disk (fractions of a second rather than
minutes to wind through a long tape) and that each item only be stored
once, no matter how many assemblies or sub-assemblies it was used in.

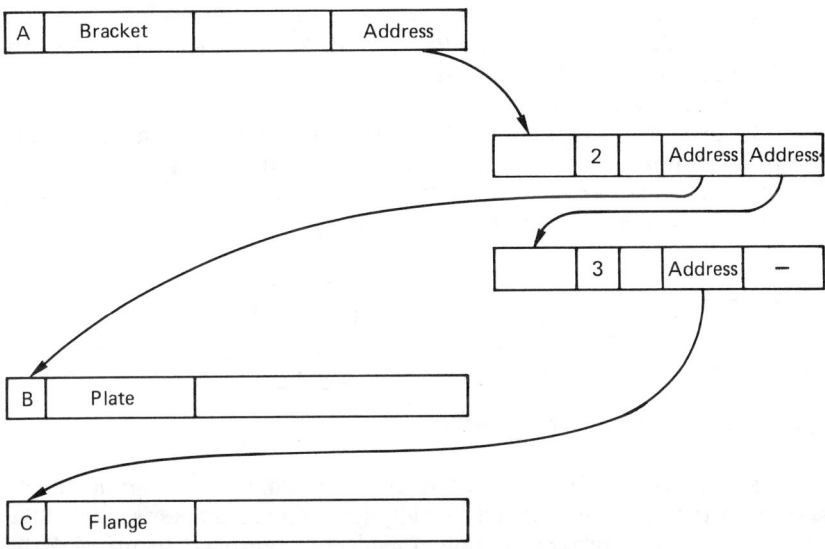

Figure 4.4 Connection by disk address between parent and component items

This (the physical organization of the data) is illustrated in *Figure 4.4*,
which gives a basis for understanding both hierarchical and relational
databases as used in manufacturing companies, but can be skipped by
those not wanting the detail at this stage.

Sub-assembly A includes 2 of item B and 3 of item C. Since A, B and C
are likely to be used in other assemblies and sub-assemblies, the fields that
contain the quantities 2 and 3 cannot be part of the record for either A, B
or C; there must be a separate record (usually called a structure record).
The basis for this is partly logical (anticipating the techniques for data
analysis that are an intrinsic part of the study of relational databases), and
partly based on the technology of disks (as already mentioned, multiple

fixed length records are preferable to varying the length of the sub-assembly record according to the number of components it has).

Thus in order to obtain the records for item A and its components, a maximum of five disk arm movements might be required. In practice, since a large block of data is read each time, it might well be less.

By the early 1970s, BOMP systems were being extended to hold routing data (Operation 10, 20 etc.) and the work centres (machines) on which these operations were performed. Routing data has similar characteristics to structure data in that each record is a relationship between the item data (e.g. item C) and the work centre (e.g. turret drill). Each item will have a varied number of routing records (or maybe none at all) and each work centre will have a very large number of operations which can be performed on it. Hence an extension of the structure in *Figure 4.4* can be developed.

One of the interesting aspects of BOMP systems was that the address field was available to the programmer, giving him the opportunity if he was sufficiently inexperienced or careless, to overwrite the address with perhaps the stock level or maybe part of the item description. When this particular assembly or routing was next used, maybe immediately or perhaps weeks later, the computer program would collapse with an error (since it could not interpret an address of 'BRAC' or 'CKET'); no reports would be available for the following day.

The next generation of such programs, by now called databases, did not allow the programmer to access the actual address fields. This made the system much more secure from error. A standard interface was provided to allow the programmer to access data without needing to know in detail how the data was stored, as shown in *Figure 4.5*.

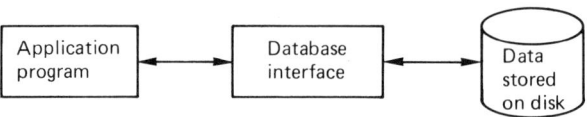

Figure 4.5 Program interface for databases

The advantages of this extra interface were numerous. Programs could access only the data they needed (the application programmer's view of the data), which made programs simpler and also improved security of the database. Hence one program could access Item A and all its components, whilst another program simultaneously accessed Item A and all the customers who had ordered it as a spare part. If both programs tried simultaneously to update the record for Item A, the interface software would handle the situation. New databases could be added (e.g. containing purchase orders) without needing to change existing application programs. Information systems specialists, probably called database administrators, could reorganize the data on the disk to reduce access time (for instance by putting commonly used records close together to reduce disk arm movement).

As computer power increases, more function is provided with databases. By setting up a separate index (i.e. a chain of disk addresses) it is possible to access any records directly. For example, all the routing records that use a particular jig or tool can be accessed directly, in addition to the normal access via item number plus operation number. This can be of value to a Jig and Tool Designer who can easily list all the items on which a jig is used, as input to a decision about making a change to it. The alternative could be hours scanning the whole database. Whether such an index has a business justification depends on how often it is used, which might be from twice a year to twice an hour, depending on the size of company and type of business. The balance of what extra function should be put on the database (at extra cost of computer power and disk space) changes continually with the reducing cost of computing. Often the users, engineers in particular, do not seem aggressive enough in demanding extra function year by year, possibly because they do not have a clear understanding of what could be available.

The technique of connecting related data records by their disk addresses, enables the storage of infinite numbers of records, including for instance purchase orders, sales orders, quotations and inquiries for each item as well as product structure, costing and routing data; it is a logical extension of *Figure 4.4* to include these extra records. Few manufacturing companies now function without such database function.

Maintaining all these addresses can still be time consuming, even when handled by the database software. Why not, as a natural extension of the move from BOMP systems to hierarchical databases, eliminate disk addresses altogether? This is one attraction of relational databases.

Relational databases

If the data in *Figure 4.4* (minus the address fields) is put into two large tables in memory, then users can access it by fitting the tables together as required. This is a relational database, and is illustrated in *Table 4.1*.

Permanent storage of the data still needs to be on disk, but (taking advantage of cheap RAM chips) it is now practical to read very large amounts of data from disk into memory to be manipulated as a relational database.

Table 4.1 Simplified view of a relational database

Part identification	Description
A	Bracket
B	Plate
C	Flange

Part identification	Operator identification	Operator description
A	10	cut to length
A	20	mill
A	30	heat treat
A	40	grind
B	10	mark off
B	20	face mill

However, data needs to be structured carefully into a database (the logical view); this is true of course for both relational and hierarchical databases. The process of structuring data is called 'normalization', and was developed on a formal basis in the early 1970s, since when the concept has been awaiting the arrival of good optimization techniques and more powerful computers.

For the main production database in manufacturing industry, the exercise of structuring the data can appear to be straightforward, because there is nearly three decades of experience in the subject. In less structured environments, the data has to be analyzed to determine how the tables (sets of tuples in the formal terminology) are constructed. If a relational database were being defined for a school, incorporating, classes, teachers, subjects, examination results, times of lessons etc, instinctive analysis does not give the answer as it may do in a manufacturing situation.

Each tuple (row) of the table must have a key that must uniquely identify that tuple. Furthermore, the key must be non-redundant, which is to say that all the data in the key is necessary for it to be unique. Taking the simple example in *Table 4.1* the table for routing data needs to have item identification as part of the identification of each row if it is to be unique. The work centre number should not be part of the key even though it would still be unique (the non-redundancy condition).

The process of normalization is done in up to three steps (first, second and third normal form). An illustration of the principles in a manufacturing context is the location of the field for the cost of a purchased item. If the cost of the item were the same for all suppliers (albeit unlikely) then cost would be part of item data in the item table. If the cost were fixed for each supplier (but different for different suppliers) then the cost would be in a tuple keyed by item number and supplier number. If however the cost changed for each purchase order, then purchase order number would have to be part of the key. If the cost changed within each purchase order depending on the date of call off, then date (or some related reference number) would also have to be part of the key.

The above example illustrates the database design depends on real world understanding; with manufacturing data it is often possible to think the subject through in these terms. Data about items, suppliers and purchase orders for example are distinct and will be held separately. The theoretical mechanics of normalization may subsequently be used as a final check on the structure, but a detailed understanding of normalization is not needed in the present context.

The question of whether to use a hierarchical or a relational database may not matter too much to the user, as long as the data needed to run the business can be brought economically to the VDU screen within a second or so. In general, hierarchical databases (more technically 'preferred-path' Data Base Management Systems, DBMS) have been more suited to the well structured, fairly static data of the business such as item data, orders, costs etc. (i.e. supporting the applications to be discussed in Chapter 5). Relational databases are better in less structured situations, an example of which would be the control of all the documents relating to an engineering change (E/C). Some engineering changes might involve only one drawing and one control document, while others might need numerous technical reports, costs of installation or maintenance, many assembly drawings, details of customer requirements, market analysis etc. In this application, a relational database would certainly be more suited; and successful experience with relational database technology is continually extending the range of applications.

Examples of the use of relational database as an integration tool will be discussed in Chapter 7 and subsequent chapters. The development of special processing hardware (vector processors and parallel processing) increases the application of relational databases).

Communications software (see below) and 'query languages' must also be considered in comparing the two types of database. Structured Query Language (SQL), initially developed by IBM, is a syntax for making inquiries into a database; the logic built into SQL is closely linked to relational database concepts. SQL has led to the development of an extensive range of 'SQL-compatible' products by different software suppliers.

Management of databases
In the evolution from BOMP to relational databases, computer systems have moved from being an optional extra for most manufacturing companies to an essential part of the business; the database must therefore be available when required and the data on it must be accurate.

The accuracy (integrity) of the data is the user's responsibility, it being generally true that not much data can be rigorously checked by a computer system (for instance, the computer cannot independently validate which product the customer ordered, or how many components were failed by Quality Control and returned for rework). However, some credibility checks are possible, such as: 'Is the order really ten times the size of the biggest order we have ever had?', 'How come more are being reworked than were in the original batch?'. Also system design can help maintain data integrity, e.g. if the customer number is entered on the VDU, then the name and address that are on the database should be displayed as a check that it is the right customer. Nevertheless, the onus is on the users to keep the data accurate; continual measures should be taken of data integrity, such as random stock checks.

The user must also have concern for the meaning of the data — for instance which of the various fields holding stock levels are used in the inventory turnover calculation. The user would usually be well advised to maintain 'data independence', and to avoid the use of meaningful part or order numbers; if, for example, order numbers of a specific format (perhaps beginning with a 9) are the ones which require heat treatment, and computer programs (and people) pick out this digit, then complications will arise if the heat treatment facilities need re-organizing — the re-organization could actually be limited by the specific logic coded ('hard wired') into the computer systems.

Unauthorized access (i.e. security) is an important issue. This ought to be manageable in a CIM system; the security requirements of financial organizations, for instance, are much more stringent than in most manufacturing companies. Security needs to be built into the database software, the operating system and also the communications software (discussed below). Professional IS skills (see below) are required, there being no point, for instance, having password security if a quickly written assembler program can read the password in the computer's memory. Similarly there is no value in building up a file that records any attempts to defeat the security system, if the file can easily be deleted.

There must be different levels of security, distinguishing between access to data, update of data, creation of new records and deletion of records. For instance, in most (but not all) manufacturing companies there would be little interest in restricting people from seeing stock records. Updating stock records (e.g. stock levels) would be limited to one set of people, whilst creation of new part numbers would be by a different set. Authorization for the acceptance of customer orders (credit checks, delivery promises etc.) would have a careful procedure, whilst sales orders and suppliers would be controlled for the obvious financial reasons, including deletion of such records. Hence security can require separate control over each field, compounded by different people maybe having responsibility for parts of systems (e.g. suppliers would be divided among the people in the purchasing department). There could therefore be a need to check, 'Is this particular person allowed to see/update this particular field, delete or create this record on this particular date?'. The date could be important because there must be a mechanism for responsibility to be devolved during holidays, illnesses etc. Personal records obviously have their own specific needs for security.

There is also a requirement to be able to recover from intermittent interruptions, such as a power cut in the middle of a program (when a data record is half updated). This requires that all updates to the database are stored as they happen, often on both disk and tape. Hence, if necessary, a new version of the database can be created from the back-up copy made last night and the duplicate copy that has been kept of all the transactions made during the day. Time becomes of essence in these matters. It is not only a question of whether recovery is possible, but also, in a network of factories, how long it will take.

Recovery also includes the need to keep a complete set of software at a remote site in case of fire or other major catastrophe. This is becoming more difficult as computer installations are more committed to teleprocessing; finding a replacement computer after a fire may not be too difficult, but provision of a set of telephone lines can take much longer.

Management of databases can be assisted by use of a 'data dictionary'. This is one of the tools of the database administrator, the function in the information systems department (see below) that has responsibility for the company's databases. A data dictionary is itself a database, but instead of holding data such as stock and cost information, it contains data about data fields and records, such as which programs access and update them, which departments have responsibility for the accuracy of the data etc. However, maintaining a data dictionary can be a significant work load.

Communications software

Communications software handles communications between different computers, i.e. computer to computer or (host) computer to terminal. The relationship of communications software to other software in computer memory was illustrated in *Figure 4.3*. If there are telephone lines between the computers and/or remote terminals, then the handling of the traffic on these lines, including the (extensive) software on control units and modems, would also be the responsibility of the communications software.

This communications technology is well established when the hardware and software are all from the same supplier (or are designed to be compatible). There are numerous companies running large (i.e. thousands) and small networks of terminals from a central computer, or from several networked computers that are running compatible software. The major suppliers have their own multi-level architectural structure for these networks (e.g. SNA from IBM and DNA from DEC, Systems Network Architecture and DEC Network Architecture respectively). Chapter 7 and subsequent chapters will be concerned with the linking together of these 'islands of computerization' when the computers are of different manufacture and running quite different software. The means of doing this will include architectures built from international standards, including OSI.

Communications software (or OnLine Transaction Processing, OLTP) enables application programs to interface efficiently with printers and VDUs (producing 'hard' and 'soft' copy respectively). To present information (e.g. stock level, item description) on different types of terminals requires a formatting process for each terminal type (i.e. explanatory text needs to be put on the screen, maybe in different languages, and the exact position on the screen of each field has to be decided). Details about these terminals should not need to be included in application programs; the additional software which does this formatting is part of the communications software.

Another function of communications software is to control security, including users' passwords. Output needs to be diverted during holidays or sickness, or if a terminal is overloaded or under repair. An interface mechanism is needed between programs (which refer to input or output terminals by a 'logical' name) and the 'physical' or actual address of a particular printer or VDU. This is similar in concept to the database interface in *Figure 4.5*.

Application software packages

There are numerous application software packages available, and they can represent a cost-effective means of getting the business benefits of computer systems quickly. The alternative is to write a one-off system (known colloquially as RYO, Roll Your Own). Some packages are designed to be modified to the user's detailed requirements (and such modification may be extensive!), whilst others are sold as complete solutions. Business benefits are of course the deciding factor, the smaller the company or business benefit, the more likely is it that the best solution will be to use a package and adapt the business process to it.

In many areas, such as CAD, In House Publishing, spread sheets, word processing and office systems, there can be very few companies who do not take standard packages.

The Information Systems department

The IS department (Information Systems, alternatively MIS for Management Information Systems or DP for Data Processing) controls and runs the central computing facilities. Whether this control extends to personal computers and shop floor networks varies from company to company. IS departments usually now charge for services rendered, rather than be carried as a general overhead charge (as has often been the practice in the past).

Historically there is tension between the IS department and its users because of the length of the software backlog, and also because the cost of the service can seem high when compared with the cost of hardware. The former complaint is not of course helped by the computer software salespeople explaining how easy the new system would be to install and how enormous the business benefits would be. This problem used to be contained because the salespeople had to work through, and could be confined within, the IS department. This, as so many other things, has been changed by the arrival of the personal computer, leading to direct contacts between users and salespeople at every industrial fair. The cost problem is in part due to the professional skill needed to keep online systems and databases available over large networks. It may be that as personal computers become more complex and are networked, users will again require professional IS support for all their computing requirements.

Another source of tension between IS and the users can be the vexing question of whether to install packages or to write software in-house. War stories exist on both sides, of home-grown software that could not be extended or even modified, and of bought-in packages that were too complex and did not work properly. This may become a significant issue in CIM; with swiftly changing business demands and major new facilities becoming available, pressure will be on the IS department to make changes quickly, whilst still ensuring that the systems are sufficiently well documented that the trainee operator can understand it. This implies increasing use of packaged software.

It is important to measure the performance of the IS department, for instance in terms of response times and system availability (e.g. what percentage of transaction response times was under 2 seconds, and for what percentage of the planned time was the system actually available). More difficult to measure (or even allocate responsibility for) is the 'up-to-dateness' of the service on offer. It is easy for the IS department to become merely a provider of a facility without being concerned how effectively it is used (as with the telephone exchange for example). Hence questions such as whether the design office should by now be using solid modelling, or whether the text processing software is out of date, can be left in a sort of no-man's land between the users and the IS department.

The issue of how well computers are used in practice leads logically to the subject of the next chapter, which looks at the flow of information in a manufacturing business.

Further reading

AMOS, S. W. (1987) *Dictionary of Electronics,* 2nd edn, London: Butterworths

AUGARTEN, S. (1985) *Bit by Bit, An Illustrated History of Computers,* London: George Allen and Unwin

CHORAFAS, D. N. and LEGG, S. J. (1988) *The Engineering Database,* London: Butterworths

CHRISTIE, B. (1985) (ed) *Human Factors of Information Technology in the Office,* Chichester: Wiley

DATE, C. J. (1986) *Introduction to Data Base Systems,* 4 ed. Reading, Mass.: Addison-Wesley

HOROWITZ, E. (1983) *Fundamentals of Programming Languages,* Berlin: Springer Verlag

MARTIN, J. (1975) *Computer Data-Base Organization,* Englewood Cliffs, NJ: Prentice-Hall

MORGAN, D. V. and BOARD, K. (1983) *An Introduction to Semiconductor Microtechnology,* Chichester: Wiley

STEVENS, W. P. (1984) *Using Structured Design, How to make Programs Simple, Changeable, Flexible and Reusable,* New York: Wiley

YOUNG, E. C. (1983) *The New Penguin Dictionary of Electronics,* London: Penguin

Chapter 5

Information flow in manufacturing

Introduction

The objective of this chapter is to provide an understanding of the information flow in a manufacturing company, some of the problems that exist in that flow, and some of the solutions in use today to try and solve those problems with computers.

The model

A model will be developed representing the information flow in a manufacturing company by considering the two main lead times — product development and material supply — and by examining the information requirements of the functions involved in those lead times. The model will then be subdivided into more familiar 'application' areas for more detailed discussion of each. The reader could use this model to evaluate how developed a company's information system is, and how that company may gain greater benefits from further computerization and integration.

The *product development lead time* is the sum of the processes involving product design, product development, manufacturing process planning, and testing. This will often include prototyping, stress analysis, model building, pre-build testing, extensive work study analyses, factory layout, assembly line balancing, flow rates and machine tool capabilities. The longer this lead time the less responsive is the company to market pressures (e.g. to requests for customized products).

The *material supply lead time* includes the supplier delivery time, the manufacturing time (foundry, bake, heat treat, paint, dry, move, machine, subassembly, final assembly), testing time, packaging and subsequent delivery time. The greater the proportion of this cycle which is non-productive, the more costly is the operation as work-in-process (WIP) and stocks increase. Typically today this non-productive element can be from 50 to 90 per cent of the total time.

The product development lead time

This lead time (*Figure 5.1*) starts with a product concept (e.g. the development or complete replacement of an existing product). This

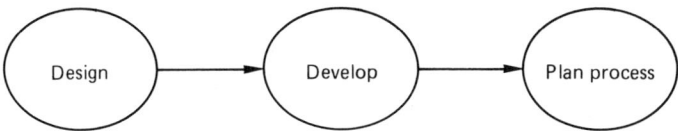

Figure 5.1 The product development lead time

concept is subsequently translated into a series of instructions to a manufacturing facility to produce that product.

Between concept and manufacture many engineering disciplines have emerged. The concept may be modelled or prototyped prior to analyzes and tests being carried out. The materials from which the product, prototype or model are made must be well understood. Mathematical techniques have evolved to enable or sometimes preclude the physical testing.

Drawings of the product, its subassemblies, assemblies and components are produced, in conjunction with the list of parts or bills of material and associated performance specifications.

These drawings, parts lists and specifications are then used as a basis for the production/manufacturing engineers to design and develop the processes and facilities for manufacture either within the same organization or by a supplier or subcontractor.

Increasing competitive pressures are forcing companies to move faster through this development cycle. This pressure results in re-design taking place simultaneously with process planning. Attempts to control these 'engineering changes' with version numbers are a major challenge in information control. Secondly, companies plan completely new manufacturing facilities very rarely. As a result, the production engineers need to know and understand the current facilities, including those of suppliers. This gives the production engineer an information assimilation task of significant proportions.

The material supply lead time

The supplier (*Figure 5.2*) provides the manufacturing company with raw material, partly-formed parts, finished parts, subassemblies or sometimes major assemblies. He will provide them against either a one off order or a scheduled delivery.

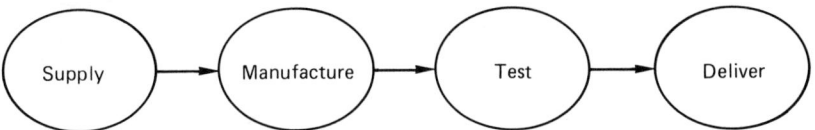

Figure 5.2 The material supply lead time

The manufacturer receives the supplied goods, and utilizes that supply in his plant, foundry, machine shop, or assembly line. At various stages between receipt from the supplier to final delivery to the customer, the manufacturer may test the material, components, assemblies or final products, either selectively or completely.

Uncertainty is a major problem. It occurs in delivery time and quality from the supplier, the results of in-plant processes, in the specification of the product and often in the minds of the customer and the sales force in terms of what the factory should be manufacturing, and to what priority.

This uncertainty, when combined with attempts at high utilization of plant and machinery, has given rise to many problems in trying to ensure that the manufacturing function has the right supply of the right parts at the right time.

Attempts to tackle these problems have had varying degrees of success. Uncertainty in sales forecasts has been addressed with statistical demand forecasting. The associated order point control, however, does not apply to many aspects of a manufacturing company's inventory. Sophisticated shop floor scheduling techniques worked very well in some large job shop environments, but often failed if feedback of what was happening on the shop floor was not available quickly enough. Attempts at optimizing the utilization of expensive machines were often a major contribution to increased work-in-process. Attempts to gain the best price from a supplier

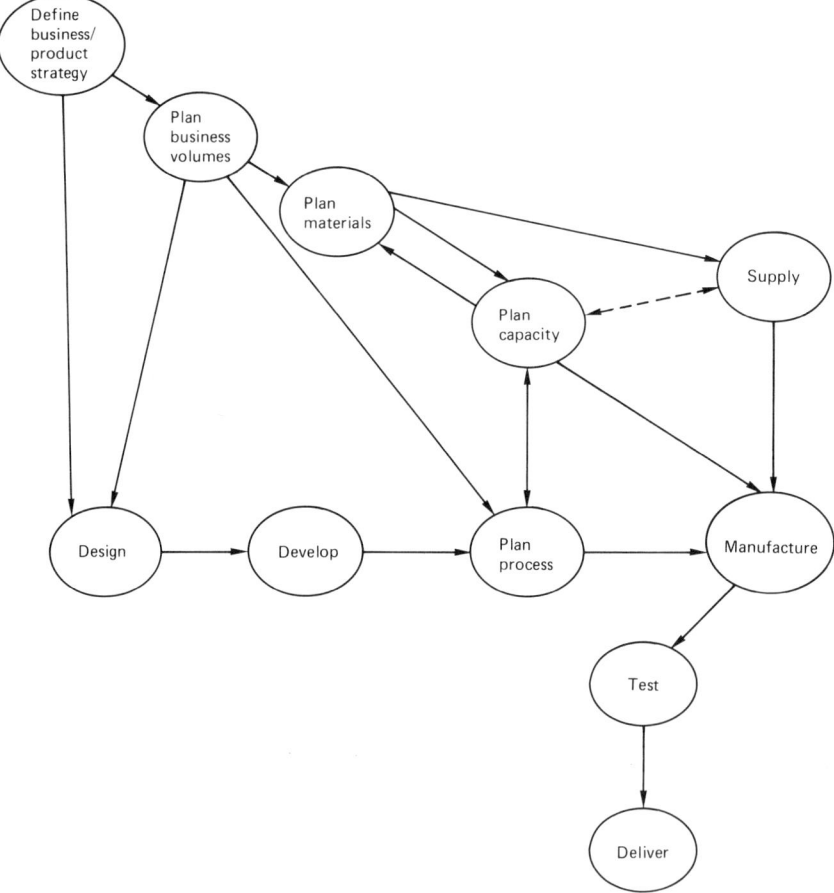

Figure 5.3 Planning, the third factor

have sometimes resulted in obsolescent inventory being acquired because communication of a product change failed to get to the purchasing department in time.

As a result, an understanding of the need to take a corporate view has emerged. This has led to company-wide production planning and control (PP&C) systems being implemented, which are very dependent on information from the design, development and process planning functions. For success the PP&C system demands that the business planning function is reflected across both lead times described above, so that the *Figure 5.3* emerges.

Completing the model

The model is developed from *Figure 5.3* and *Figure 5.4* by identifying some of the elements of information that are exchanged between the different corporate functions. This cannot be exhaustive and the reader may wish to augment from personal experience.

The examples shown in *Table 5.1* are listed to illustrate the principle that information produced in one department is used in other departments, and therefore a coherent use of this information across all functions of a corporation is essential.

Table 5.1

Originating function	Information	Utilizing functions
Design	Product structure	Development Process planning Master production schedule planning Materials planning Manufacturing Purchasing Technical publications Sales
Process planning	Routing sheet	Master production schedule planning Capacity planning Manufacturing
Purchasing	Open purchase orders	Materials planning Capacity planning Manufacturing
Master production schedule planning	Master production schedule	Purchasing Manufacturing Sales
Sales	Customer orders	Master production schedule planning Materials planning Capacity planning Manufacturing Purchasing
Manufacturing	Operation details	Materials planning Capacity planning Sales

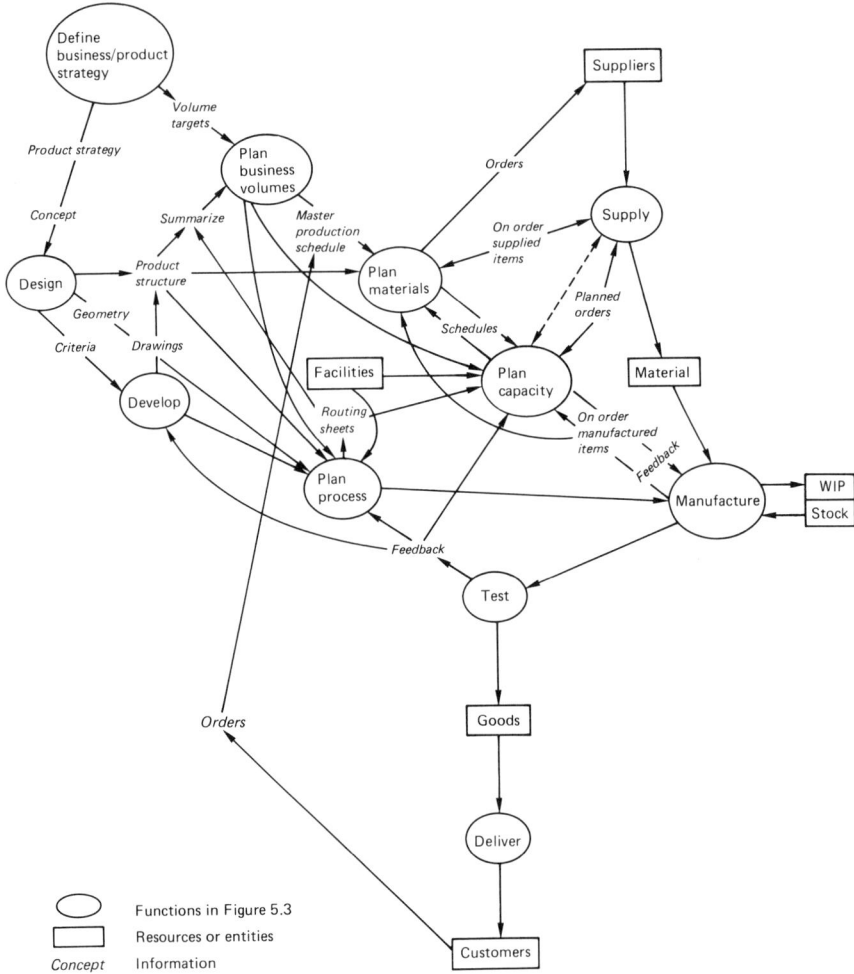

Figure 5.4 The information flow model

Functional information requirements

The *business planning* function starts the information flow by establishing the company objectives in terms of product volume and profit objectives, in conjunction with a basic knowledge of the capabilities of the plant. These global corporate goals are passed to the design function, the material and capacity planning functions and the sales function, for example. These corporate goals may include statements of direction like quality product image, low-cost manufacture, short customer delivery time, etc. (as discussed in Chapter 2) which will become part of the affected department's objectives.

The *design* function, with the product image and cost targets, will call on material specifications and costs, as well as its own technical knowledge in developing the product. A knowledge of competition will also be necessary. Depending on the industry, the design may now be handed to various development or test functions. The design information will take the form, at this stage, of drawings — two and three dimensional, exploded views, details, cross sections — of material lists, of product structure information, and performance criteria. Today, in some CAD systems, many of the views of this design are integrated to ensure that as the product design evolves, the views of that product, the stresses on the product and the dimensions, for example, are kept in synchronization. The use of CAD systems is described in more detail in Chapter 8. In a more traditional environment, however, most of these drawings are produced and maintained separately. The drawing information is also passed to the process planners, to the material planners and sometimes to the suppliers. The sales department will need their view on the specifications and appearance of the product for sales negotiations, and purchasing will need to know of any new materials or volumes of existing ones which may have to be re-negotiated. The technical publications function will need the drawings to produce sales literature, technical and customer manuals, and specification sheets.

The *development* function will use the design information to build prototypes for testing and to build solid or mathematical models for analysis. The results of these tests have to be recorded and stored for future reference, as well as used to influence the design. The gathering and analysis of this information is a major task in some environments. The prototypes built at this stage are often used to prepare sales literature, as well as test cases for the processes of manufacture. In some industries, with the aid of CAD, these different uses of the base geometry of the design of a product are bringing these different departments closer together.

The *process planning* function, provided with drawings and parts lists, need information about the company's existing facilities to design the manufacturing process. The product design or the cost targets may result in new equipment or techniques being employed. The process planner therefore needs additional information about new machines and techniques. Information about facilities includes speeds, feeds, capacities, unit costs, batch size constraints for tools and maintenance, set-up times, etc. To enable the use of NC machines or robots, the process planner will use the drawing to establish the machine instructions.

The *master production schedule planning* function (discussed in more detail below) will utilize the business plans along with a summarized product structure and a resource requirements statement to simulate the feasibility of the business plan and to produce a master production schedule. Part of the simulation may include various alternative product, process or volume strategies combined with knowledge of external market pressures. This will have established that the plan for the product, along with other products, fits the company targets, that there is sufficient capacity in global terms, and long lead time items have been identified.

The *materials planning* function, with the master production schedule and the product structure information, will produce the material schedules for purchasing and the factory. To obtain the correct schedules for these two sources will require information about current on hand inventory, as well as factory-based and supplier-based orders.

The *capacity planning* function will need the material plan (produced by the material planning function), the process sheets and the facilities available in terms of shifts, number of machines, etc. This function will produce information about over and under loaded facilities, late or early deliveries and, in the shorter term, a work-to-list for the factory floor. For these statements to reflect reality as closely as possible will require (as with material planning) an up-to-date version of the state of orders already started in the factory.

The *manufacturing* function will need information about which orders to start next, the material content and process plans for those orders, and the tools and facilities required to execute them. The material content from the design point of view may need to be rearranged to produce kitting lists or stores issue lists or assembly line sequences. This function will need access to stocking information for their kitting process. The order and its operations will have been prioritized to reflect the current version of the master schedule (reflecting current customer order due dates). Changes in specification to an order from the customer or from engineering will need to be communicated as soon as possible to the manufacturing function. The time spent at the different facilities, and by the manpower on the shop floor, has to be recorded for costing and subsequent profitability analysis. Measurement of the success rate of the material conversion processes is assessed to ensure improvement, where necessary, in the design of either the product or the process.

The *purchasing* function uses the material plan as a basis for negotiation with the supplier. Purchasing will need a drawing and the specifications for the item to be purchased. Information about each supplier, including their terms and conditions, delivery and quality histories, discounts and current order book need to be available. Purchasing will need information about the quality and quantity of material delivered to authorize payment of the suppliers' invoices.

The *sales* function, having received sales targets as their part of the business plan, need to understand delivery or 'available to promise' schedules for negotiating sales with their customers. They will need to monitor actual sales against their targets. They may be involved very closely in the design or specification process for the product being ordered, in which case they will need very precise technical specifications, including drawings and models of the finished products; they may need glossy sales literature with schematic and three dimensional views, or photographs of the sales items. All this information originates in the design function, and may have had to be redrawn in a technical publication function or may have been obtained by photographing a mock-up or model. The sales function will also need information about those customers who are behind in their payments.

The *technical publication* function will use the design data to produce sales leaflets, technical specification sheets, maintenance manuals and customer operation manuals. They may also combine selected aspects of this with the company business plans and results to produce both internal and external facts sheets as well as the annual report.

The *accounting* function will be the recipient of information about costs, sales, time spent, and material consumed as the basis of business profitability assessment and business planning.

Fundamental to all these functions will be information about the structure of the company's products.

Product structure information

Information about the material content of the finished product is important from a planning and control perspective; it is also complex to manage from both a company organization and an information systems viewpoint. One factor is that the departments in an organization who use the information have developed their own ways of rearranging it to suit their own needs, for example:

- a parts list on a design drawing is a simple list of those components in the drawing,
- an assembly presented as an exploded view of its parts has that list rearranged according to the resultant layout, with a cross referencing to a sequence number on the drawing,
- where used lists — in which all the assemblies containing a particular component are referenced to that component — are used by accounting to assess the impact of price changes,
- sales options and variants,
- manufacturing engineering's need to rearrange 'as designed' into 'as manufactured', when an assembly (e.g. a braking system) is only completed during final assembly,
- modifications to the product or material specification once launched, with the necessary release control,
- material schedulers' need for a level by level view of the product content to minimize inventory by reflecting stages of manufacture,
- expeditors' kitting lists for marshalling parts prior to assembly. This list may have to exclude certain items such as fastenings which are available as free issue in the assembly area.

In 'one off' projects that are designed and built only once, or where a simple product has only one level of manufacture, the problems are much reduced.

One common element of today's solutions to the above problems is a corporate database of this information. The departments responsible for that information are the ones to maintain it; while those who only need to view it can have it presented in the format that is most productive to them. This chapter applies to all manufacturing industries where the 'bill of material' may be referred to as a parts list, a recipe, a cutting list, a specification sheet, a formulation, etc. The term 'product structure' will be used for this information.

The *product* is what a company ships to its customers. This may be a raw material, a component, a major sub-assembly or assembly, or packaging; or it may be a product, such as a car or a can of food, that is immediately useable, and complete in its own right. This product may be made from one or more assemblies, subassemblies, packaging, parts, chemicals or materials; many of those in turn being made from raw materials, other chemicals, paper or glass or metal, or yet more parts and subassemblies.

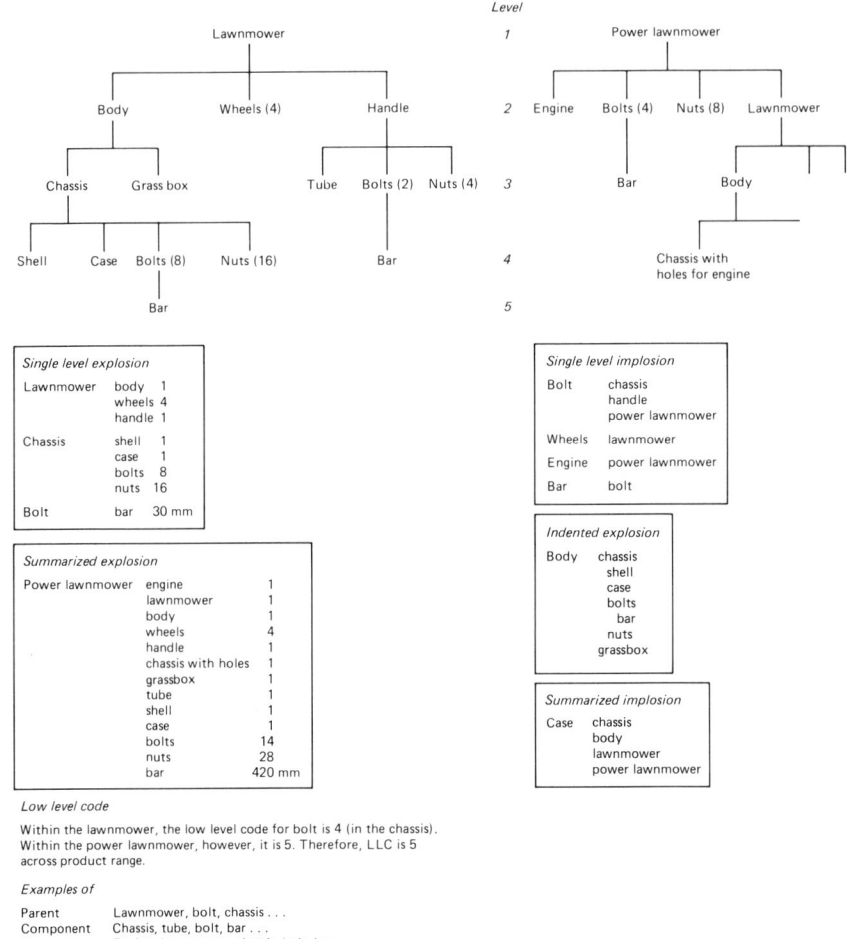

Figure 5.5 Product structure

The starting point for all is some form of raw material. The relationships between these various stages of manufacture are defined via the *product structure* database, which is also the source of where used information.

In the following section the engineering manufacturing industry will be used as the mainstream example. This is largely because it tends to represent the more complex end of the spectrum in this product structure discussion, and so solutions developed here can usually be applied elsewhere.

The basic information

The item information must be separated from the product structure information, as discussed in Chapter 4.

Item information would include, as well as the identifying number:

- description,
- unit of measure,
- costs: standard, actual and deviations,
- stock information: on-hand, on order, order quantity, allocated quantities,
- lead times: complete or in terms of set-up and run times per piece,
- production or procurement plan,
- routing and drawing numbers,
- engineering change or modification history and status,
- any other information unique to that item.

Product structure information would contain:

- quantity of the component in this parent (this may be expressed in units, length, weight or volume depending on the characteristics of the component and the industry),
- engineering change number, status and effectivity date,
- lead time elements specific to this relationship, such as a component which is introduced part way through the manufacturing process for this parent,
- operation number at which this component is used,
- option or variant information such as percentage usage.

With this separation between item information and product structure information, database technology is then utilized (as described in Chapter 4) to enable the appropriate storage and retrieval. Six basic views of the product structure data — known generically as one level, summarized or indented explosions or implosions (see *Figure 5.5*) — can provide the information needed, for example, for:

- a summarized component list for purchasing (summarized explosion excluding intermediate parents),
- a single level parts list for kitting (one level explosion excluding those components which are line stock),
- a complete cost build-up of a major assembly from the costs of the lowest level components,
- the basis for MRP (level by level explosion).

An additional requirement is the need to prevent the 'used-on-itself' syndrome, where an item can inadvertently contain itself as a component at that or a lower level in the structure; another is the need to know, during the MRP sequence of item planning and explosion, when all dependent requirements for any component have been generated. Both of these can be achieved through a device known variously as low level code (LLC), lowest level usage indicator or assembly level. This identifies the lowest level across all products at which the component in question is utilized in a product structure (see *Figure 5.5*).

In addition to LLC, other database validity checking needs to be carried out. For example, if an item is to be deleted or archived, because it is now 'obsolete', the database should not contain stock of that item, nor should the item be a valid component in a current parent; nor should orders be outstanding in the factory or with a supplier. Individual companies will have their own additional examples of the need for validity rules.

Additional considerations

Sales versus production identity
The identity of a product in a company may be different from the identity of the product to a customer. There is often a similar problem between manufacturing and engineering, where an item is often referred to by its drawing number.

Options and features
The final product specifications are often tailored either to a specific customer or a specific application. Where additional function is provided, the term option is often used, e.g. air conditioning in a car; and where a range of a function is available, the term feature or variant is often used, e.g. the engine size. With either of these, associated structure changes need to be recorded and processed, without having a complete duplicate set of many product structure records. Decision tables have been used to try and ensure that invalid combinations of features (from, for example, an engineering or space point of view), cannot be ordered.

Artificial structures
It is often a requirement to impose some artificial structure into a product structure database. For Master Production Schedule Planning (MPSP), for example, a very much simpler view of a product or product group may provide sufficient information. Within the production process itself, it may be convenient to group certain components together for planning and stock picking purposes, although they are never formed into a proper assembly. These devices are referred to variously as M-numbers (M for manufacturing), phantoms or pseudo items.

Engineering change control
There are two main requirements. One is the need to be able to plan the phasing-in to the factory of one component, or group of components, to replace another; the timing of this may be forced on the situation for safety or economic reasons. The other is the need to gather together and consider all the information that normally would be taken into account when deciding on a change of this kind. The first of these needs to be able to be implemented automatically through the information presented to the planning system, thereby preventing the replenishment of obsolete parts as soon as possible. The second can be addressed by the stock, sales, on-order and where-used information available on the databases described above.

Master production schedule planning

All companies need a long-term plan. That part of it used in production planning is known as the Master Production Schedule (MPS), and it provides communication from senior executive management to production. The information requirements for the Master Production Schedule

Planning (MPSP) can be simplified into two main headings, the product plan and the summarized resource requirements.

The product plan should reflect the overall business objectives and plan. In doing this the shape and content will vary from business to business, but should contain sales forecasts, including launches and promotions, customer orders or schedules, an indication of what is available to promise to customers, major assemblies or products and/or options and variants.

The summarized resource requirements may consist of a simplified product structure including only 'significant' assemblies or long lead time components; it may include a time-phased resource statement for the critical resources in the factory. One of those resources may be machines, one money, one labour. In a business more geared to design-to-order the critical resource may well be the design engineers or the process planners.

The product plan

It is essential to have the product plan in an easily digestible form, concentrating on the important elements. What elements are most important will depend very much on the business being planned.

In a steady, high volume business one of the decisions reflected into the product plan will be whether to build to stock or only against customer orders; and if against customer orders how far ahead should the customer feel committed to either specific orders or longer term schedules. The tendency of retail outlets and major manufacturing companies to be moving towards requiring 'just-in-time' (JIT) deliveries emphasizes the need for this even more.

In a lower volume, option-oriented business where the base product is modified towards the end of the manufacturing cycle to a specific customer order the product plan may well be much more like build-to-stock, from a planning point of view, using the sales intelligence to guide the scheduling and forecasting of options.

In the build-to-order business for defence or aerospace or larger customization business, there is much more emphasis on long term estimating, and associated risk in holding or developing production facilities. As competition and cost-cutting pressures increase in these sectors, however, the need to be able to predict a shorter contracted delivery date is putting pressure on those suppliers to have a generic product plan to be used as the base; this is often augmented by spares or consumables manufacture as a buffer.

Whatever the business type, there are two functions which are common. These are the forecasting of products, major assemblies, options, spares or consumables; and balancing actual customer orders or schedules against that forecast.

Forecasting
The only predictable aspect of a forecast is that it will never be right. However, techniques are continually being developed to apply a mechanistic approach to the interpretation of the past combined with best estimates of the future to provide a quantitative view of the likely sales of a

product in its life cycle. Subsequent careful monitoring of the 'accuracy' of this forecast can then be used to modify subsequent plans.

One simple model for intrinsic forecasting sales is known as exponential smoothing, whereby sales statistics, gathered over time and grouped into consecutive time periods of equal length, are weighted by a factor to predict the sales in following periods. This prediction can then be modified by knowledge of, for example, seasonal influences or product promotions, to arrive at 'the forecast'. A better version of this model from the company's perspective would be to use 'demand' rather than 'sales', but success in assessing demand and feeding it into such a model has eluded most companies.

In the fashion industry, however, capitalizing on that demand can be crucial to the success of the business. Many of those companies are moving towards mobile inventories where they can supply to a retail outlet within hours, thereby ensuring that they have little or no surplus inventory at the end of the season while providing a high level of customer service.

The limitations of this intrinsic forecasting, which only looks at past performance as a basis for the future, are recognized. Planning functions in companies today therefore often use extrinsic forecasting models based on demographic data or government gathered statistics of, for example, house build starts or predictions of disposable incomes and life styles, as a basis for their plans.

Another aspect of forecasting volumes for MPSP is the expected consumption of spares or consumables, both of which are predictable due to the design of the product. One knows, for example, that a new oil filter is required say every 6000 miles and that the average motorist drives 10 000 miles per year.

On the spares side, many components have a previously defined life expectancy, and this can be used in conjunction with anticipated usage and model life to predict spares requirements. This prediction can then be monitored against actual orders to establish a more realistic build programme, as well as informing engineering.

No matter which aspects of forecasting are deemed useful or necessary, there are two inherent questions. First, how good is the model? This requires careful and consistent monitoring. Second, how much of the forecasting can be automated? One person cannot be expected to monitor thousands of lines in the same way that a computer can. Using the computer model's monitoring mechanism to highlight unusual deviations for the human planner to investigate makes good use of both resources.

Balancing customer orders against the forecast
This requirement exists when the build programme has to be established well in advance of customer orders being obtained, and yet it is not planned to build for stock. This is often caused by long lead time components or materials; or increasing competitive pressures.

It is necessary to distinguish in the build programme that quantity of the schedule that is already allocated against specific customer orders; when the customer request date is modified, the reallocation of the build programme can take place to satisfy other customers' demands. By continually having this information available, the sales forces can have confidence in quoting delivery times when negotiating with customers, and production can anticipate their load with some certainty.

As the business links between supplier and customer develop, and the build programme reflects the supply schedule requirements, the supplier can use it as a basis for his own plans.

Summarized resource requirements

The summarized resource requirements data consists of a time phased reflection of the major components, options and variants, production facilities, money or people, held in a hierarchical manner like the product structure database discussed above. The time phasing, via multi-levels, can be offset simply, or accumulated down through the levels required.

Judgments have to be made with regard to which elements should be present in this simplified model of the product plan's content. It could be a complete model of the business, with the final output being the cash flow and profit and loss statement. More realistically, its output will highlight the need for extra cash funding, extra subcontracting or extra factory capacity requirements that might result in purchase of a new machine tool or extending a factory.

The point is that the planner, in trying to arrive at a reasonable production schedule, needs to identify the constraints and take the appropriate steps to circumvent them.

As a simple example, consider a final product which has three major options, one of which needs significant design engineer involvement. To help promote sales towards the end of the product's life a special campaign will be mounted. The MPS planner would draw up the following resource list:

	Months prior to delivery
Major assembly A	4
Major assembly B	6
Option X, 10% occurrence	6
Option Y, 25% occurrence	6
Option Z, 65% occurrence	6
Option X, Engineering 20 hours	10
Promotion Department 100 hours	18 months from now

Each of the major assemblies in turn will have production requirements as their 'resource', since both of them occur in other product lines as well.

MPSP functioning

Given the product plan and the interrelated summarized resource requirements, the calculations performed by MPSP are straightforward. They are very similar to those of MRP, and capacity planning (see below), and in some cases companies would include the inventory and capacity commitments from the operational systems in the short-term part of the plan.

The results of such a calculation will be to present management with a summarized profile of requirements for those critical resources covered by the plan. Once these results are reviewed the management team can apply 'what if' questions and gain the real benefits.

If the results of the MPSP calculation are acceptable, no more may have to be done. However, management may wish to assess the impact of trying to increase sales by rerunning the calculation to see if that identifies any constraints or bottlenecks.

More often it will be necessary to re-align the plan to try and remove any identified constraints. This may be possible by adjusting, for example, parameters of sales percentages, lead times or promotions, and redoing the calculation.

Used in this way, one would anticipate not using MPSP every day in most companies. It would be used, however, when a new product was in development as a prelude to factory development, supplier relationship and cash planning. It could also be used to evaluate the success or failure of a product if its life cycle was, say, 25% shorter or longer than expected; or to assess the benefits and costs of replacing part or all of a factory or workshop, or replacing certain problem components with a more expensive but more reliable one; or to assess some of the benefits of moving from a flow line machine shop to a flexible manufacturing cell; or to assess the introduction of robotic assembly, welding or painting operations. More and more it will be used to help companies plan with their suppliers the effects of moving to JIT.

Material planning

The objective of material planning is to establish material availability in the factory to satisfy the build programme with minimal inventory holding and therefore low costs. The build programme in the form of the master production schedule represents a consolidation of customer orders and planned production volumes to reflect the business plans. Appropriate capacity is assumed to be available through the use of MPSP.

The basic challenge in material availability is to have the right part at the right time. This means placing orders with suppliers and the factory itself in sufficient time to receive the goods when they are needed. Two basic techniques have been used to achieve this: order point control, and material requirements planning.

Order point control

The order point control model is illustrated in *Figure 5.6.*

This shows the stock level over time for an item, given reasonably continuous usage of the item with some fluctuation in that usage, and also in the replenishment lead time (X,Y,Z) possibly improved (Y) due to expediting of the supplier and (Z) due to the supplier having excess stock himself. In this model the fundamental assumptions are:

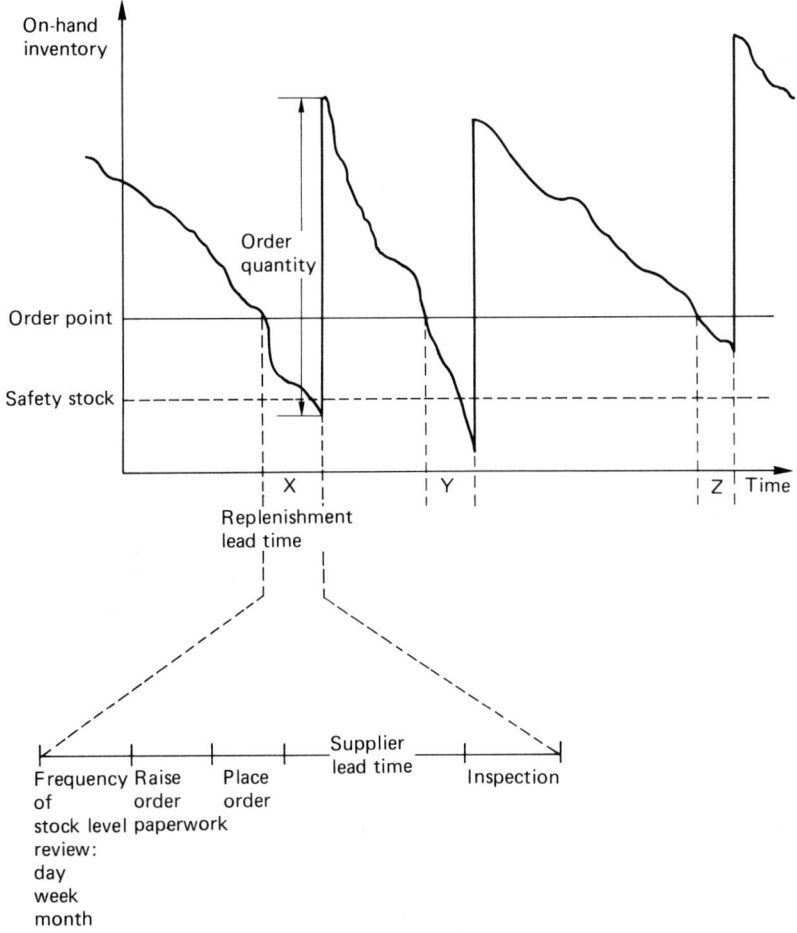

Figure 5.6 Order point control model

- usage/demand is continuous, and reasonably predictable,
- the degree of unpredictability can be measured statistically, and can be used to set safety stock levels, and reorder level.

Safety stock is inventory held to provide cover for unreliability in the replenishment lead time from the supplier, or the anticipated usage being greater than expected. Reorder level is the inventory level set to trigger the reordering, and will be based on the expected demand during the supply lead time.

The above assumptions are often not valid for items in a manufacturing company where batching of sales, customizing options, seasonal dependency and engineering changes may cause discontinuity in demand. Resultant excess and obsolescent stocks are often difficult to avoid.

Much of this problem has been addressed by utilizing MRP which lays out a schedule for re-ordering based upon the master production schedule. The results on inventory holdings are illustrated in *Figure 5.7*.

Order point control model: order triggered by stock level

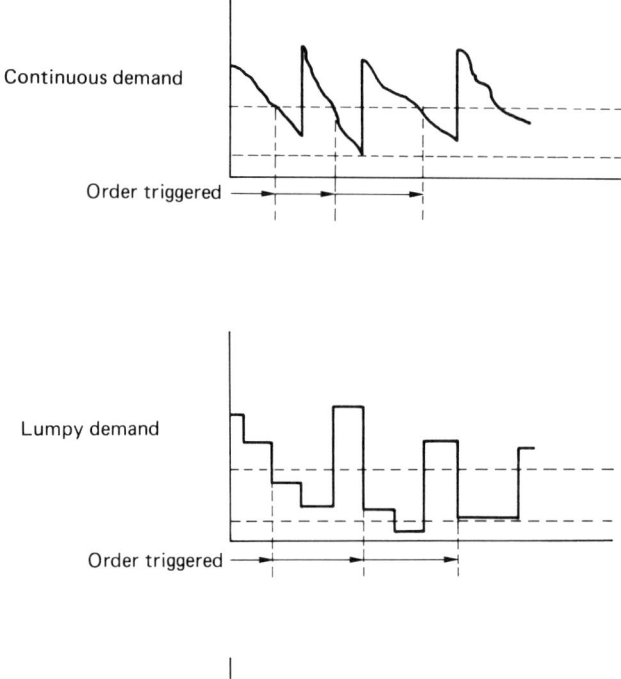

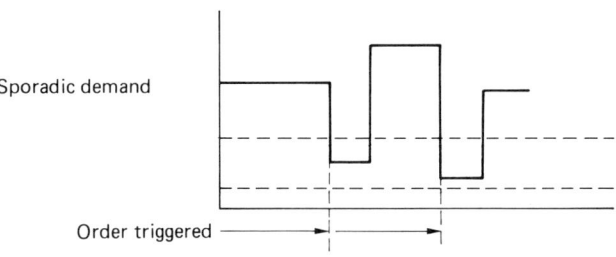

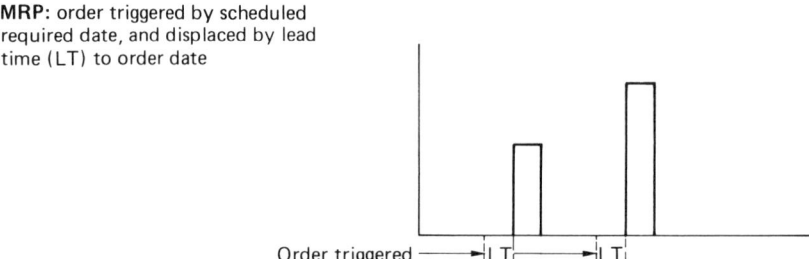

MRP: order triggered by scheduled required date, and displaced by lead time (LT) to order date

Figure 5.7 Inventory holdings: MRP versus order point. The inventory holdings (area under curves) is much less in the lowest diagram. Although exaggerated with sporadic demand, the benefits exist in all cases, especially when combined with the anticipation of engineering change control

Material requirements planning

As shown in *Figure 5.8*, the basic information requirements for an MRP system are:

- master production schedule,
- on hand stock figures,
- purchased and manufactured order status,
- replenishment rules:
  order quantity
  lead time
  safety stock,
- product structure.

The results of the planning are suggested orders, known as planned orders, for each item being planned and changes to current outstanding or open orders.

Information requirements

Master production schedule   The master production schedule consists of all items to be shipped from the factory: finished goods, kits, spares or inter-factory orders, or some combination of these. For MRP the schedule should exist far enough into the future to cover the complete manufacturing and procurement lead times (e.g. 9–18 months). This is necessary not only to ensure the availability of supply, but to have a base for planning capacity.

On-hand stock figures   An item may be stocked across several locations in a factory or company. Stock exists often at the assembly line as well as in stores. There may be production and spares stocks. There may be inventory physically marshalled prior to shortage chasing and expediting. All these sources of inventory must be in the on-hand figure.

Accuracy of these stock figures should be a top priority of any implementation of MRP. It is not uncommon for stock record accuracies to be as low as 90% and that this is only established at the annual stock check time. However, where well-executed and controlled cycle counting systems and on-line data collection and reporting systems exist, figures nearer 100% are being achieved. (Cycle counting is a system to ensure that stocks are counted regularly and compared with the record of inventory.)

Order status   MRP produces a series of planned orders and recommended actions to fulfil the MPS. These planned orders are the responsibility of the MRP system and until the status is changed by the planner, MRP assumes that the order can be changed in quantity or date as the MPS changes. The parts planner may decide to confirm or release an order to the factory or to a supplier. This changing status of an order must be recorded and allowed for in subsequent replanning.

With purchased parts, an MRP system may prompt buyers to check with the supplier that all is in order ahead of time; with manufactured parts, any monitoring and feedback of operation completion dates with any associated scrap reporting from the shop floor could be modelled to predict a completion time. Any discrepancies with the required date could be highlighted.

Process

MPS for finished goods and spares

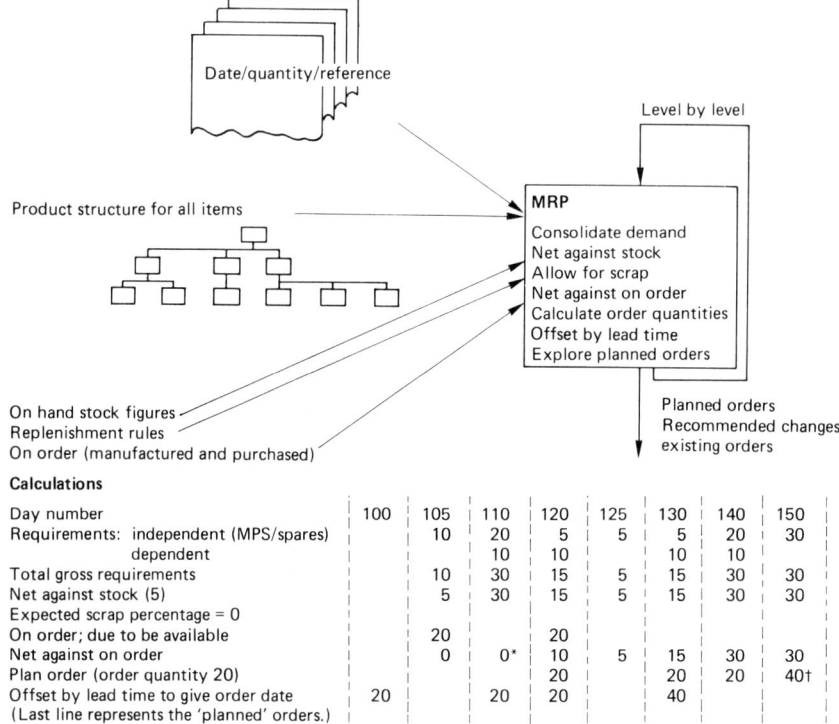

Calculations

Day number		100	105	110	120	125	130	140	150
Requirements: independent (MPS/spares)			10	20	5	5	5	20	30
dependent				10	10		10	10	
Total gross requirements			10	30	15	5	15	30	30
Net against stock (5)			5	30	15	5	15	30	30
Expected scrap percentage = 0									
On order; due to be available			20		20				
Net against on order			0	0*	10	5	15	30	30
Plan order (order quantity 20)					20		20	20	40†
Offset by lead time to give order date		20		20	20		40		
(Last line represents the 'planned' orders.)									

*action necessary to bring order due day 120 forward
†place two simultaneous orders

Lead-time considerations

Planning horizon for MRP should be long enough to include supply and manufacture lead times:

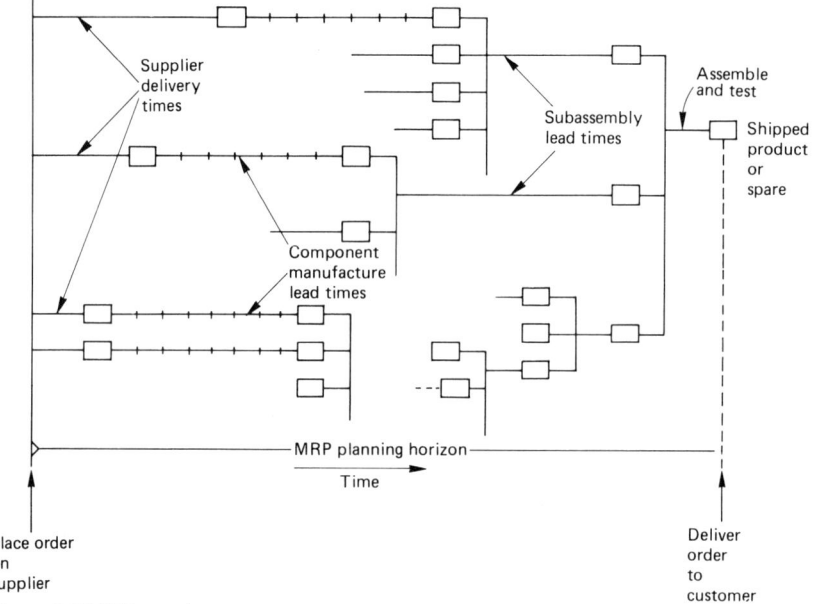

Figure 5.8 MRP overview

Replenishment rules The quantity to be ordered is determined via the following techniques:

- order to schedule: the order quantity is exactly equal to the requirement quantity,
- order by time period: the order quantity is the sum of all requirements in a day, week or month,
- economic order quantity: the order quantity is calculated to balance the costs of ordering against the cost of carrying stock, and is based on assumptions about annual usage of the item,
- part period balancing and least unit cost: variations on the Economic Order Quantity (EOQ) algorithm to allow for the calculated requirements in MRP,

and these above may all be modified by considerations of:

- maximum,
- minimum,
- multiple,
- discount breakpoints from suppliers.

The order should be placed in time for availability by the required date and this lead time may be stated simply in terms of a number of days or weeks. For purchased parts time has to be allowed for paperwork processing and delivery. For internally manufactured parts lead times can often be calculated and will vary with the quantity to be manufactured. Information about queue time, set-up time and run time can be used to calculate this lead time.

Product structure The essential piece of information about the product structure for materials planning is the 'quantity per'. MRP should also use the effectivity date for a pending engineering change and an offset in lead time if not all components of a parent are required at the same operation.

MRP calculations
MRP takes the independent demand for each master production schedule item and breaks down that plan using the product structure data to form a dependent element of the build programme for each component. If some of the build programme inputs are spares, it is very likely that they appear as components somewhere in the end product's structure, and since one of the principles in MRP is ensuring that prior to planning an item all dependent and independent requirements for it have been gathered together, the spares items would not be planned till this was complete. This is achieved via the low-level code as discussed above under product structure information.

For each item, once all such requirements for it have been gathered, item planning can take place. This takes the following steps:

- consolidate requirements as a gross plan,
- net off any stock against this gross plan and allow for safety stock,
- allow for scrap in the manufacturing process,
- net off any purchased or manufactured orders currently outstanding,

- apply any order quantity calculation rules to establish quantities and due dates for suggested/planned orders,
- offset this date by the procurement or manufacturing lead time to calculate order release date.

This results in a series of planned or suggested orders.

This series is then exploded via the product structure to calculate the dates and requirement quantities for the component(s) of this item.

Enabling the plan

During MRP several conditions may be brought to the attention of the planner. These will be in addition to providing the planned orders for those items for which the planner has responsibility.

These conditions may include:

- order should be expedited due to change in plan,
- order should be delayed due to change in plan,
- order should be increased/decreased,
- orders which are past due,
- situations where the quantity ordered needs to be greater than the maximum of the rules.

The planner can then review these against the plan and, in conjunction with his own local knowledge, take the appropriate action.

In many companies MRP is used monthly or weekly to produce the 'current' plan. Two major problems with this total gross replanning have emerged. The first is the sheer volume of information with which the planner has to cope; the second is that often the plan is overtaken by events and becomes obsolete. To assist with the former, the MRP system will compare the planned orders produced this time with those produced the previous time and highlight only those which have changed significantly. Coping with the second concern has resulted, in conjunction with the evolution of computer technology, in the concept of 'net change MRP'. The principal concept of net change MRP is that replanning should take place only when there are changes in the data on which the plan was based.

Net change MRP

If the master production schedule has changed dramatically, it is very likely that all concerned will know and start to take the necessary action. If, however, the change is less significant, such as an item's stock record was found to be incorrect on a regular cycle count there will be a need to have the computer trigger the communication. Then it would be useful to run that change through those parts of the plan where it occurred and review the resultant changes to the plans for any dependent components. Filters could be built-in to smooth any over-reaction to small perturbations, while absorbing the actual change accurately.

As an example of this, take the case of replacing a component X by a component Y for a limited period due to material scarcity of X. If the resultant change to Y's schedule was outwith its manufacturing lead time and the lead time of its raw material, then there would be no need to do anything precipitately. The normal planning procedure would take place, turning a planned order (for the new quantity) into a committed order either with the supplier or the factory. If, however, the result required the expediting of either the raw material or the component's manufacture, it would be essential to highlight this for action immediately.

Capacity planning

The basic information requirements for capacity planning are:

- orders currently being worked on in the factory,
- orders anticipated (planned orders from MRP),
- routing sheet,
- facilities information.

The output from a capacity planning system will be two fold. The first is the anticipated load for each work centre or facility in the factory. The second is an indication of where there are problems in achieving the planned dates established by MRP. The capacity planning system itself may well be sophisticated enough to take the necessary rescheduling decisions, but often the best it can sensibly be expected to do is to recommend.

Orders

The two sets of orders — planned and released — are the basis for planning the workload, and need to be current. The MRP system provides that currency for the planned orders and up-to-date data collection provides it for those already released to the factory.

The data held about each of these sets of orders is different in one respect. The released orders will have the actual set of operations necessary to carry out this particular order, several of which may have been completed, whereas the planned orders will only have the due date for the order as a whole, and will use a standard routing as a basis for capacity planning.

Routing sheet

This generic term for the set of assembly instructions, machining or fabrication operations, including heat treating, painting, cooling, drying, as well as mixing, packaging and probably inspection or test, is as fundamental to capacity planning as is the product structure to the MRP system. The set of instructions, as with product structure, is subject to change, both formal and informal. Often on-the-job practices are used to overcome the process planner's route under the pressure of pragmatic production. These evolutions of the routing sheet should be reflected into the information used for planning.

For each operation, the system needs to have the identity of the work centre to be utilized, the expected time required of that work centre (both set-up and run times) and any information on queue and move time both before and/or after this operation. In more sophisticated systems, there will also be an indication of whether this operation can be split between two or more machines to help throughput time, or whether this operation can be overlapped with the preceding or succeeding operation, and possibly whether an alternate facility or work centre can be used in the event of an overload occurring on the proposed resource. The run time may be expressed in standard units and have to be converted to real units, or it may be expressed in time per 100 rather than per one off.

Set-up time for this operation, any need to refresh the set-up after producing a specific quantity, or any specific waiting time such as cooling or drying may well be associated with a particular operation.

The sequence of operations on the routing sheet usually reflect constraints on the sequencing of operations in practice. However, sometimes operations can be done 'out of sequence'. If this is prevalent in a factory the routing sheet information should hold that information to enable alternative sequencing to be planned.

Facilities information

By a facility is meant any resource whose capacity should be considered in the plan. In many factories a work centre such as 'lathes' or 'drills' is considered to be the correct level of detail required for this, but this may only be properly considered if that work centre's machines are all of a similar type and capability, or at least capable of performing similar operations. This is becoming less true as flexible machining cells are evolving where several machines are necessary to provide the right combination of facilities to be able to manufacture a family of parts, or a group of parts whose manufacturing operations are similar. In most cases it is necessary to itemize on the facilities database all machines albeit grouped into work centres.

It may also be necessary to itemize the assembly stations on an assembly line, or the assembly line as a whole. It may be, however, that the assembly line has been designed to cope with a certain volume of output, reflected in the master production schedule in the first place, and therefore it would not be necessary to have this in the capacity planning exercise.

The information required for capacity planning would include available capacity and any queue and move times associated with that particular facility.

The capacity available may fluctuate over time, with the number of shifts, planned maintenance and anticipated changes in production output. There may be a regular shift pattern during a week rather than a day, and public holidays and annual shutdowns must also be catered for.

Capacity planning process

The basic function of capacity planning is simple. By starting with a due date for an order, which consists of several operations requiring the resources of identified work centres, and allowing for queue and move time between operations, it is an easy algorithm to backschedule in time to identify when the particular work centre is going to be required. By accumulating all such work centre resource requirements, usually referred to as loads, in some suitable time bucket such as a shift, a day or a week for example, an overall picture of the anticipated load at all work centres may be obtained (*Figure 5.9*).

Complications begin to emerge in two situations. One is if there is not enough capacity to cope with the anticipated load; the other is if, in backscheduling, it is discovered that a particular operation, or indeed a whole order, should already have started! Neither of these is uncommon,

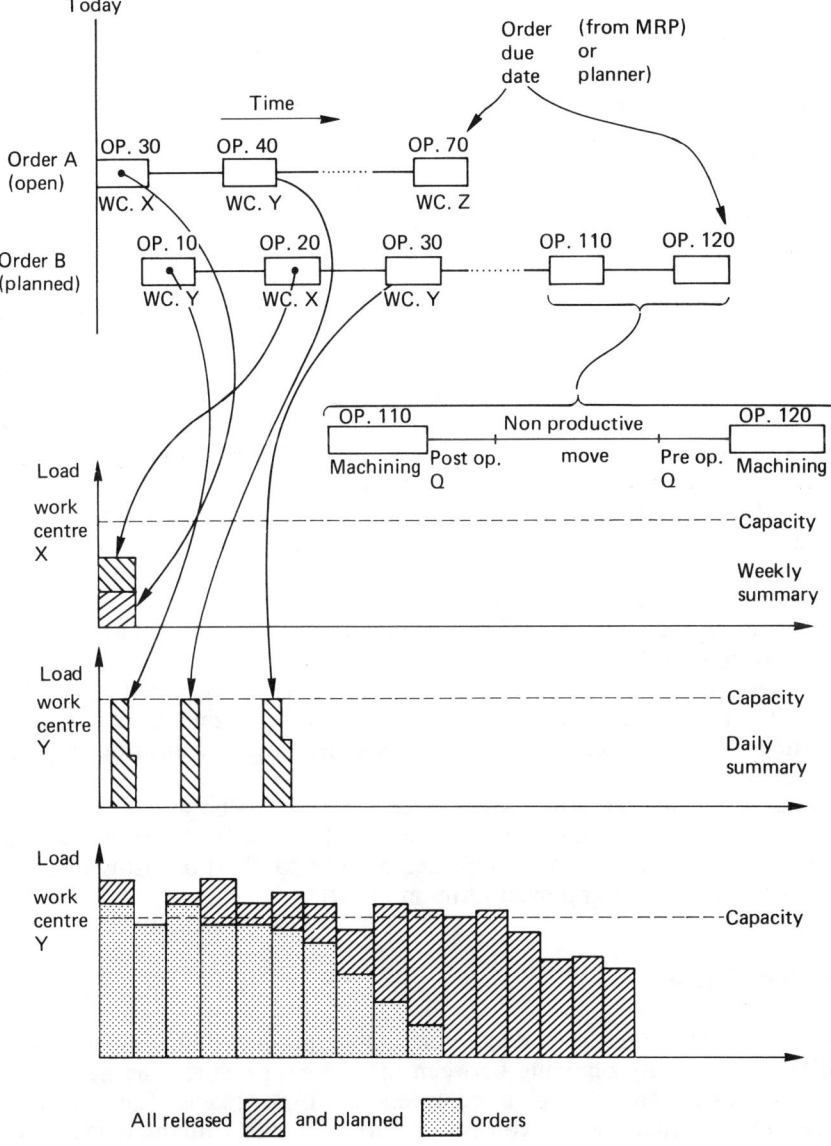

Figure 5.9 Capacity planning overview

which is why capacity planning systems are designed to help alleviate both of these problems.

Where an overload on one or more facilities occurs there is a need to establish priorities, prior to re-scheduling and providing information about orders which will be late, or may have to start early. The major objective is to satisfy the MRP dates, but many other priority algorithms have been developed. Lateness combined with due date is an obvious one. Some systems allow an external priority to be assigned, or consideration to be given to added value or eventual cost, or to the level of the component in the product structure hierarchy or to the elapsed time still to run.

Examination of lead times in manufacturing companies shows that up to 90% of that lead time is often non-productive. In trying to keep to planned dates it has been possible to shorten the overall lead time by the following means. The productive element can be shortened by overlapping succeeding operations, by utilizing a machine with a faster throughput or by splitting an operation across several machines if they are available. The non-productive element — almost always squeezed by expeditors in shortage chasing mode — lends itself more readily to planned time savings. First move time can be shortened, but probably not down to zero. Secondly, queue and wait time can be shortened down to some predetermined limit. In all of these attempts to keep to the schedule, a good capacity planning system will not only apply these tools, it will also let the scheduler or planner know what it has done.

Often, however, it will not be possible to achieve the planned dates and the material plan will have to be changed.

Capacity planning usually starts by loading backward from a due date to establish the normal latest start time. Then an order is loaded forward from the earliest start date to establish the earliest finish time. Both of these will be done without consideration of capacity constraints and this is known as loading to infinite capacity. From these calculations lateness and slack can be determined. These aspects can then be used to prioritize and hence rearrange workload to cater for both the overload problem and the 'should have started' problem.

Loads will then be recalculated to a finite capacity based upon the established priorities, where necessary either squeezing lead times or starting some work early to free up resources later for an order with a higher priority.

Finally, no matter how sophisticated the rules to be applied are, the system will recognize that some facilities must be overloaded or the orders will be late. The load is often represented by a barchart or histogram and more often today in graphical form on a VDU.

Order release

Given that MPSP has provided a 'reasonable' product build plan and that MRP and capacity planning between them have provided the necessary scheduled start dates for planned orders, the order release function now takes over. Order release consists of three functions for manufacturing orders. First there is a need to review the availability of the components or raw material for that order and allocate that inventory; secondly, the operations have to be scheduled to provide a realistic guide to the foreman for his work; and thirdly shop paperwork has to be produced either on paper or electronically to reflect the results of the preceding two functions. For purchased orders, the purchasing department may wish to review planned orders prior to committing them to suppliers.

Shop orders

Order release takes place as late as possible to keep inventory and work-in-progress down. With a computer system it is possible to assess material availability without physically kitting or withdrawing from stores. There is the additional advantage that the most recent versions of the material plan, the product structure and the routing will be used.

Inventory allocation

The planned order start date plan should provide the logical priority sequence for inventory allocation since these dates reflect both the material plan and the capacity plan. However many companies still feel the need for external priority sequencing prior to inventory allocation.

The process of inventory allocation reviews each planned order due for release in a predetermined number of days into the immediate future. This will usually be set at two to five days, although sometimes for major assemblies it may go further out. For each planned order, inventory will be allocated when all components are available, this allocation being debited from the on hand on withdrawal from stores or consumption on the assembly line or as a result of final completion of the order initiating what is known as back-flushing a debit of inventory. This review and allocation is achieved via the product structure, which will be at its most current level. In some customized or customer options businesses, the product structure may be a 'special' at this stage.

If components or raw materials are short, a decision has to be made as to whether or not a particular order should be released. If an order is released with only some of the components allocated, the availability of those components to other orders is unnecessarily constrained. On the other hand, releasing such an order and producing an associated shortage report may result in the higher priority order being completed on time.

After allocation of the inventory, it is necessary to produce an electronic or paper version of the routing for shop order tracking on the shop floor, and to facilitate operation sequencing. For each operation on the routing database, a record will be created on a shop order database reflecting this order's quantities and current version of the routing. As with the material content, this may be customized to reflect any decisions made earlier about modifications made as a result of customizing or capacity constraint decisions. This shop order database subsequently acts as the repository for status information on partial or successful completion of operations as the order progresses through the shop; or the consumption of material against the planned inventory allocations. As such it is the basis for the MRP and capacity planning or inventory availability and resource requirements information. This feedback function to the database is often referred to as a closed loop system.

Operation sequencing

Once released orders have been added to the shop orders database, it may be necessary to perform some short-term shop floor scheduling to produce an up-to-date version of a work-to-list.

Often achieved during the capacity planning process, but more often rerun with a much shorter horizon, and to finite capacity, this function usually looks only two or three weeks ahead. Several different techniques have evolved to make this task more effective. Most of them are variations of the procedures described in the capacity planning section, but one known as input/output control addresses itself specifically to the practical day to day aspects of operation sequencing. The objective of this technique is to balance the flow of work through the shop floor while recognizing the fact that certain work centres or machines are most likely to be the constraints on the rate of flow with which the shop floor can cope. This is achieved by guiding the planners towards not releasing an order till reasonable expectation exists that it will flow through the shop floor efficiently. Implementation of this technique requires the recognition that bottlenecks do exist, and the better implementations include the historical build-up of performance data to help in this process.

The work-to-list generated can be then used by the foreman to plan set-ups, check for tooling and whether his feeder department is on schedule or not.

The credibility of this part of the system depends crucially on the accuracy and speed of the feedback from the shop floor. This feedback requirement is difficult since events on the shop floor happen quickly compared with, for example, product structure or routing change activity. The introduction of Programmable Logic Controllers (PLCs) (see Chapter 4) in shop floor equipment is beginning to make this feedback more effective.

Purchase orders

Purchasing should be involved with the design and manufacturing planning departments early enough in the product development phase to have arranged quotations, if not contracts, with appropriate suppliers. They may have also established alternative sources for items, with associated terms and conditions on price, quantity and delivery time. All of this information could be held on a supplier database to enable monitoring and analyses of performance in terms of promised versus achieved levels of quality and delivery time dependability, for example.

This database, built up over many years, is the handbook of the purchasing department so that if emergencies in supply arise, they can guide the company through events such as suppliers going out of business, large customer orders needing extra supplies, a unique customizing requirement, or a machine breakdown giving rise to the need to buy rather than manufacture.

Whether this database is available or not, purchased orders have to be placed to fulfil the planned orders resulting from MRP. In some cases the orders will be one-off to probe a new supplier's credibility. In other cases the orders will reflect only a minor adjustment to an already contracted call-off schedule, where a supplier will deliver as requested within time limits based on an annual or monthly requirement quantity estimated by MPSP. If neither of these applies there may be a need, based upon the planned schedule, to negotiate directly with the supplier at that time. The buyer may then be discussing discount quantities against requirements, balancing delivery schedules against inventory holdings, or negotiating penalty clauses for late delivery or poor quality. This is where the database enables the review of the historical performance of suppliers.

The important aspect of the above with regard to order release is that the latest planned orders, and adjustments to released orders both accurately reflect the most up-to-date version of the build plan, the product structure and stock figures and are available quickly to the purchasing department. This is also a fundamental prerequisite to just-in-time delivery contracts, both from the supplier's point of view and that of the factory ordering the goods.

Once the order is released to the supplier, the database must be updated to reflect this status so that subsequent planning runs are cognizant of those decisions made by the purchasing department.

Shop floor data collection

Collection, analysis and feedback of the status of activities on the shop floor is crucial in measuring the success of a manufacturing business. Although referred to as shop floor data collection, and often implemented for limited requirements such as labour reporting for payroll and costing and operation completion for operation scheduling or work-in-progress (WIP) calculations, this key operation in a manufacturing company can be used as a source of information to control the business as a whole. Used effectively, the information can update the status of orders and therefore:

- answer customer enquiries on delivery schedules,
- result directly in stock savings,
- allow expeditors to work more effectively,
- enable foremen to make better decisions on work sequencing.

It can also be used to monitor quality. In turn this could trigger:

- design towards redesign or modification,
- product engineering towards process changes or tool redesign,
- maintenance department towards more frequent maintenance of a particular machine,
- retraining of an operator.

This need for data collection on the shop floor has been satisfied historically in many different ways.

One of the earliest was the time sheet, when each operator, via his foreman, had an authorized labour operation completion ticket that was used as a basis of payment of the man who had performed the operation. It was recognized that this was not necessarily useful as a basis for material or capacity planning, as frequently the time sheet reflected work that the operator wished to be paid for rather than all the work done.

To improve the accuracy of the information being gathered, separate systems evolved with punched cards, magnetic striped cards and, more recently, bar coding which could be used to report completion of operations independently of the payroll and costing process, but updating work in progress and shop order databases. The identification of the order and the operation number were read 'automatically', thus reducing errors, while quantity completed and scrap had to be keyed in manually, with the foreman or someone other than the operator authorizing the data entered. These trends in data collection are reflected in computer technology appearing specially designed for the shop floor, with ruggedized equipment and specially designed consoles and keyboards.

With the need for this information to be more up-to-date through the use of MRP and capacity planning systems, so has the need to record time for set-up of a machine prior to an operation, as well as the start and stop times of the operation itself. Although this information is used to update a shop order database, it is potentially of value to the production engineers and process planners to enable them to monitor the success of the estimating in the manufacturing processes they have designed. Prompted by the need to monitor the quality of the process, as well as the time estimate, they have initiated component or product testing as a separate operation. Taking place after the manufacturing process is complete and the analysis of the results being even later means that any faults in the process of manufacture will not be corrected for some time after scrap has been produced.

In parallel with these developments, numerical control (NC) and direct numerical control (DNC) equipment is being increasingly used to direct the operations of the machines on the shop floor, from robot assembly and spraying through cutting and welding to Automated Guided Vehicles (AGVs) and automatic warehousing systems. The direct linking of these PLCs to feedback mechanisms for counting parts, reading bar coded components for batch identification, measurement devices for checking quality either as it is happening or immediately after, will become more common place as the use of computing technology evolves.

This will allow a more direct link in the control of the manufacturing business more like process plants. The ability to measure frequently, to analyze 'on-line' and therefore raise any warning flags very quickly will contribute significantly to improving the quality and reducing the cost of the items being manufactured. By comparing the analyzers' results with the product specifications immediately much scrap will be avoided. Faulty machines, designs, processes or materials can be spotted almost immediately when further wasted added value will be avoided.

In summary, the data collection function on the shop floor is now capable of monitoring the manufacturing business directly within the factory in such a short turnaround time that many of the costly errors of yesterday will be avoided.

Summary

In the applications described above, there is a major element of commonality in the information used. The flow of this common information between departments in a paper oriented system contributes significantly to the length of both the product development and material supply lead times. It is the length of these lead times which contributes to the inability to be responsive to change (in product development) and to the high cost of carrying inventory and work-in-progress (in material supply).

Today's technology allows the information flow to be replaced by information access. Through the use of database and online computing systems the information can be maintained by the right people and can be accessed by the responsible people so that better decisions can be made more quickly based on up-to-date and available information.

Thus different functions can benefit both themselves and their company by having quicker feedback about the effects of their functions. For example:

- sales can have reliable information to give promise dates to customers,
- design engineering will learn more quickly that they have designed a product which is difficult to manufacture,
- process planning will discover earlier in the manufacturing life of a product that a particular operation is proving unsound,
- material planning can reduce obsolescent stock by having more immediate information about product structure changes,
- manufacturing can reduce scrap by being more quickly aware of machines going out of tolerance.

The company can benefit through better quality, more responsive design, shorter lead times and reduced costs of inventory, work-in-progress, scrap and obsolescence.

The attainment of these benefits will need much careful planning. The education required by all members of the company to accept their responsibility for their data, and to action their exception reports produced by these computer systems must not be underestimated. The methodology is, however, proven and the benefits are substantial.

Further reading

GUNN, T. G. (1981) *Computer Applications in Manufacturing,* New York: Industrial Press

HALL, R. (1983) *Zero Inventories,* Homewood, Illinois: Dow-Jones Irwin

LOCKYER, K. (1983) *Production Management,* London: Pitman

PLOSSL, G. (1985) *Production and Inventory Control, Principles and Techniques.* Englewood Cliffs: Prentice-Hall

VOLLMANN, T. E., BERRY, W. and WHYBARK, D. C. (1984) *Manufacturing Planning and Control Systems,* Homewood, Ill.: Dow-Jones Irwin

WALLACE, T. F. (1985) *MRPII: Making it Happen,* Essex Junction, Vt: Oliver Wight Limited Productions

WIGHT, O. W. (1974) *Production and Inventory Management in the Computer Age,* Epping, Essex: Gower

Chapter 6

Simulation

People making decisions about CIM need to use simulation as part of their planning. Simulation can be a valuable tool for design assurance, and can save costly mistakes particularly in large capital-intensive projects. Between 2% and 4% of investment seems to be an accepted yardstick cost. Historically, however, the use of simulation has sometimes not been properly planned, with consequent disappointment in the results. Including simulation skills in CIM planning is a first essential, backed up with a proper awareness on the part of all concerned of what to expect from simulation. Successful application of simulation demands good understanding and good management, and these are the subjects of the two halves of this chapter.

Understanding simulation

In this chapter the term simulation will be confined to what is known as 'discrete event simulation', in which *time* is the primary quantity modelled. Modelling normally involves the compression of time, and the main objectives of conducting simulations in CIM can be summarized as increasing production and decreasing inventory, or for short TPUT (i.e. throughput) up, and WIP (work in progress) down.

A simple demonstration using pencil and paper shows how the utilization and build-up of queues at a machine can be studied, given a series of intervals between the successive arrival times of parts being delivered, called for short 'inter-arrival times' (IATs), together with the corresponding process or 'service' times which the parts experience at the machine. Suppose that the data is as follows:

IATs   0 1 3 5 1 2   service times   4 2 2 4 3 2

where in each case the data is continued by repeating the same cycle of six values. Make up a table with column headings as in *Table 6.1*. The entries in the EVENT column are marked with 'ar' for an arrival event and 'eos' for an end of service event.

The entries could be continued indefinitely. Suppose that a halt is made at this point. Some basic statistics about queueing and congestion at the machines are already calculable:

Table 6.1 Pencil and paper simulation

Clock	Event	Queue length (incl. in service)	Next arrival	End of service	Time not in use	Cumulative wait time (incl. in service)
0	1ar	1	1	4	0	0
1	2ar	2	4	4	0	1
4	1eos	1	4	6	0	7
4	3ar	2	9	6	0	7
6	2eos	1	9	8	0	11
8	3eos	0	9	0	0	13
9	4ar	1	10	13	1	13
10	5ar	2	12	13	1	14
12	6ar	3	12	13	1	18
12	7ar	4	13	13	1	18
13	4eos	3	13	16	1	22

Number of completions (TPUT)	4
Maximum queue length (WIP)	4
Average wait time (incl. service)	$22/4 = 5.5$
Average wait time (excl. service)	$(22-(13-1))/4 = 2.5$
Server utilization	$12/13 = 0.92$

Types of simulation
The simplest way in which a computer can be used for simulation is the straightforward automation of calculations, such as those above, which can be accomplished with a tool as simple as a humble BASIC interpreter. As more sophisticated languages are brought to bear, the modeller is relieved of some of the burden of programming by the introduction of more complex computing ideas such as lists and procedures, and also of the tasks of collecting the most commonly wanted statistics. As computer sophistication grows still further, so does the specialization in concepts. Typical abstractions created by the designers of simulation languages are 'transactions' and 'facilities'. In the case of parts moving through a series of machines, the parts themselves might naturally be modelled as transactions seeking the services of machines which are modelled as facilities. On the other hand it is equally possible to model the machines as never-ending transactions which seek the 'service' of a generic part playing the role of a facility.

It is becoming increasingly popular to conceive the units of manufacturing systems in terms of networks, using this term in the sense of directed graphs rather than linked clusters of computers. *Figure 6.1* illustrates a simple manufacturing system which is modelled as a queueing network.

Input to a computing system demands that the network diagram be converted into a character representation. The package called SIMULAP (see Dangelmaier and Bachers 1985) does this as in *Table 6.2*.

Representations of manufacturing systems in this style are however too crude to accommodate the full complexities and interactions involved in real automated systems. In the mid 1970s a form of dynamic network called

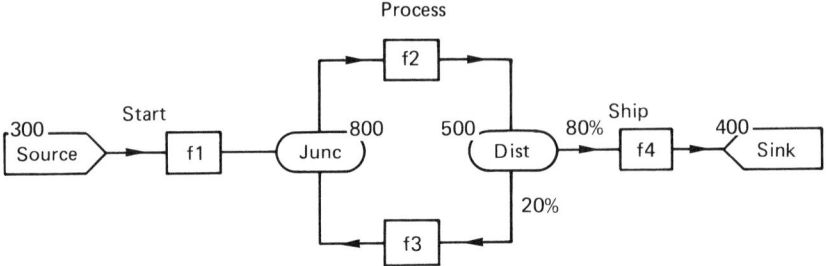

Figure 6.1 Network model in style of SIMULAP. f1 stands for facility 1, etc. Processing times at the facilities are: f1, 10; f2, 8; f3, 20; f4, 5

Table 6.2 Simplified SIMULAP model

Facilities	No.	Process time	Capacity	Successor	Predecessor
	1	10	1	800	300
	2	8	1	500	800
	3	20	1	800	500
	4	5	1	400	500
Sources	No.	Successor	IAT	Start-time	Number
	300	1	10	1	50
Sinks	No.	Predecessor			
	400	4			
Distribution nodes	No.	Algorithm*	Successor nodes	Percentages	
	500	1	4 3	80 20	
Junction nodes	No.	Algorithm*	Predecessor nodes		
	800	3	1 3		

* A variety of algorithms are programmed within the package which relate to the details of how competing units of flow are dealt with at distribution and junction nodes.

a Petri net was devised, which is effectively a network like that above, but overlaid with a system of tokens which are initially placed on some of the nodes, and that subsequently jump about through the network according to a set of so-called 'firing rules', thereby giving the network a continuing dynamic momentum.

Petri nets were originally devised by computer science theorists to describe the workings of large complex computer programs such as compilers, and research in this area has focused on analysis rather than modelling. It is only in the mid 1980s that they have been considered by manufacturing people as a basis for describing their sorts of systems, and some experimental implementations of Petri net based systems have been reported (see e.g. Abraham and Sciomachen (1986)). One has to note the excessive length of the descriptions which are required to specify even a very simple system completely.

Another pair of matching high level concepts is 'process/entity', which emphasizes the contrast between processes as *active* agents, and entities as *passive* objects on which the processes operate. This gives an alternative and quite different way of modelling the parts flowing through the machines, and as with the transaction/facility scheme, the modelling can be carried out in either of two ways. The more natural way is to model the machines as the processes (active) operating on the parts (passive), but it would be equally valid to model the parts as the processes conducting their affairs by operating on passive machines.

The general point to emphasize is that modelling presupposes deliberate considered choices between a range of concepts in the abstract programming world, and a collection of physical objects — machines, parts, and material handling systems — that populate the real world. The modeller makes his choices in two stages. The first choice is that of the *range* of concepts, a choice that is likely to be dictated by cost and availability of programming tools within the organization. Subject to this choice, the second stage is the matching of computer concepts to physical reality, and it is at this point that modelling techniques and modelling skill come strongly to bear. There are no absolute right and wrong ways, rather is it a case of more and less elegant ways.

Since modelling techniques are highly dependent on the computing system, it is not possible here to do more than point out some generalities that apply across all simulation systems. Techniques in detail properly form part of the documentation of individual systems, which are now so numerous that it seems unlikely that it will ever be economic to produce general volumes on core techniques. There are nevertheless a few problems that must be repeatedly remodelled in different systems, and thus deserve a few comments here.

The conveyor problem   In a transaction/facility system a conveyor with infinite capacity can be modelled straightforwardly as a multi-server facility. Real conveyors however generally have the effect of spacing out the items that they carry, and thus imposing a continuous delay on the system of which they are a part. This means modelling the conveyor as a *series* of facilities, whose number is equal to the capacity of the conveyor, a procedure that often leads to extended run times. In a process/entity system, it is more easily managed by keeping tallies of items in transit and doing comparison arithmetic on times. At the end of one process, an entity is marked with a future time, which represents the earliest time at which it can receive the attention of the next process. Before the latter starts to operate on it, the current time and the marked time are compared, and if the latter is ahead of the former, then the process goes into a suspended state until the corresponding variables become equal.

More general material handling systems   In situations where the handling system possesses intelligence say in the form of AGVs, transaction/facility systems require the forward movement of a transaction to depend on the simultaneous availability of more than one facility, which has always been a tricky area in GPSS (General Purpose Simulation System) and similar languages. In a process/entity system, an AGV is accommodated as a handling process interacting with the machine processes.

Breakdown and maintenance   These considerations are often ignored in the initial construction of a model, but sooner or later authenticity is bound to force their inclusion. Their introduction involves greater speculation than that of the regular performance of machines since the latter can usually be reasonably precisely predicted in advance by engineers, whereas the former requires a period of sustained production to obtain some realistic 'history'.

Breakdowns in a process/entity system can be either events that happen following a random decision at an appropriate point within the cycle of a machine process or alternatively breakdown processes can be introduced, perhaps with a separate breakdown process matching each machine process.

Distributions

The concept of a statistical distribution is one that simulation specialists are apt to take too readily for granted in discussions with their clients. Although this concept is usually well enough understood by those whose training or experience contains a statistical element, it can be a new and relatively bewildering concept for those whose background education has been in other disciplines, and in these cases time is well spent explaining it. Interestingly, a statistical distribution is itself a model, although of a quite different kind from a simulation model. The main quantities whose distributional properties are important are inter-arrival and service times. The simplest and most widely-used distributions in simulation studies are uniform, normal, beta, exponential, and Erlang.

The exponential distribution is intimately connected with Poisson processes which are characteristic of systems where individuals are alike but their arrivals into the system are independent of each other. An example is pallets arriving on an uncongested conveyor system, with a mean arrival rate which is constant over time. Examples of what are *not* Poisson processes are pallets on a congested system, or pallets that arrive as clusters by lorry-loads.

The next stage in developing simulation is to seek means of obtaining random values for quantities such as IATs whose distributions are among those listed above. This leads to the requirement of *random number generators*. A very simple random number generator for the exponential distribution can be devised on a pocket calculator. If m is the mean inter-arrival rate, it is easy to show mathematically that the formula

$$-m\log_e r$$

where r is a fraction drawn randomly from the uniform range $(0,1)$, returns a random value from the exponential distribution. Where the client's background is not statistical, it is worth spending time to emphasize the difference between the above formula and the equation

$$y = (1/m)\exp(x/m)$$

which describes the probability density function of the distribution.

Making exponential and similar simplifying assumptions sometimes allows the use of analytical techniques. For example, a model with a single service centre employing n servers in which both IAT and service times are exponentially distributed is known as an M/M/n system following a notation invented by Kendall (1953). (The Ms stand for Markov whose name is associated with the early mathematical studies of such systems.)

In this case the following formulae apply (see Page (1972) for proofs). Suppose

a = mean inter-arrival rate (i.e. 1/mean inter-arrival time)
m = mean service rate (i.e. 1/mean service time)
c = number of servers
a/m is called the traffic intensity, and estimates the minimum number of servers necessary to avoid congestion

$u = a/cm$ is the average server utilization.

Write

w = waiting time, i.e. including service
q = queueing time, i.e. excluding service
s = service time

so that

$$w = q + s$$

Theory dictates that under these circumstances, the long-run average values of w, q, and s are:

$$\frac{1}{m(1-u)}, \quad \frac{u}{m(1-u)}, \quad \frac{1}{m}$$

respectively, and the average numbers in each state are:

$$\frac{u}{(1-u)}, \quad \frac{u^2}{(1-u)}, \quad u$$

respectively. The rth percentiles for w and q are respectively:

$$\frac{1}{m(1-u)} \times \ln \frac{100}{(100-r)} \quad \text{and} \quad \frac{1}{m(1-u)} \times \ln \frac{100u}{(100-r)}$$

Suppose for example that a vendor makes a mean number of 10 deliveries per 8-hour day, each delivery requiring an average of 30 minutes use of the loading dock. Overall dock utilization (u) is clearly 62.5%, yet the contractor complains that his lorries often wait over an hour to unload, and it rarely takes less than 90 minutes to turn round a truck.

Writing $m = 1/30$, $u = 5/8$, the average queueing time (excluding service) is 50 minutes, and the average number in the queue is 25/24. The 75th and 90th percentiles of w using the formula above are 111 and 184 minutes respectively, so that in spite of an overall 62% utilization, the contractor's demands for a second dock are not altogether unjustified.

Whereas most of the time such analyses will confirm commonsense expectations, every so often a surprise of this nature occurs. The difficulty with analytic methods is that if there are more than one or two service centres, or if the distributions of IATs and service times lie outside a relatively restricted set, then the mathematics becomes too formidable to handle. Further there is a danger of getting too carried away with the attractions of such simple distributions, and using them indiscriminately where the theoretical basis for their use is scant. For example, in studying the arrival of lorries at a loading bay, it could be better to use empirically known patterns of arrival, e.g. with morning or afternoon peaks, which may be far removed from any of the standard distributions.

Variability

However enthusiastic the exposition given by the simulation specialist, the client is unlikely to remember the details of many statistical distributions, nor are subtle differences likely to be of much practical consequence. In simulation the finer distinctions of statistical analyses are rarely of great account — what the client wants is advice where massive differences in cost and performance hang on design decisions. As Conway *et al.* (1986) say of such effects:

'If you can't see it, forget it!'

Consequently it is important to have a single concept which embodies the essence of variability in a way that is common across all distributions. Such a quantity is the *coefficient of variation* which is the ratio of standard deviation to mean. Thus variability can be perceived as a quantity that varies on a scale from 0 to 1 with 0 representing total uniformity, and 1 representing the variability associated with Poisson processes and the exponential distribution. This is about the greatest degree of variability likely to be met in practice, although there is no technical reason why the coefficient of variation (henceforward known as COV) should not exceed 1 numerically.

The potentially damaging effects of variability may not always be as well understood by engineers as they should be. Suppose that we model a situation shown in *Figure 6.2*.

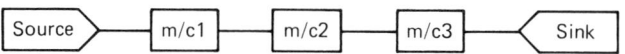

Figure 6.2 Machines in series with no buffers

Parts are generated at a source and are processed in sequence by each of three machines with no intermediate buffering. Suppose further that the mean IAT and the mean processing time for each machine is 5 time units and that the simulation is run for 10 000 time units, so that the target or 'ideal' TPUT is 2000. The graph in *Figure 6.3* shows the pattern in which simulated performance falls short of the target.

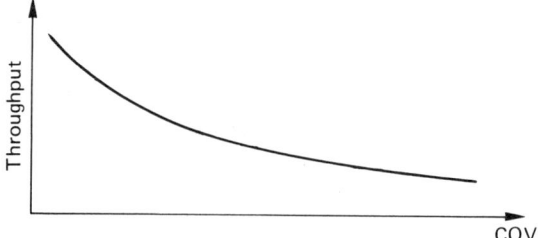

COV

Figure 6.3 Relationship between throughput (TPUT) and coefficient of variation (COV)

As the use of simulation becomes more commonplace in an organization, so it should become progressively more natural for all concerned in CIM planning to become acquainted with the idea of COV and make automatic subjective assessments of its value. Where the ultimate aim is to increase TPUT, much of this planning will turn out in practice to be a matter of how to counteract variability by buffering.

Communication and presentation

The activity of industrial simulation today is much bound up with colour graphics on a computer, and no simulation system or package can become popular unless it can be shown to produce effective and meaningful graphics. Arguably some systems are sold on the basis of the glossiness of the graphics rather than on the quality of the modelling they allow. A useful ABCD of necessary 'cosmetic' functions ancillary to simulation is:

ANNOTATED OUTPUT, i.e. traces of individual runs;
BUSINESS GRAPHICS, charts of throughputs, utilizations, etc.;
CONVERSATIONAL MENUS, offering the user the opportunity to change parameters just before the start of each run;
DYNAMIC ANIMATION, providing the visual evidence of program execution.

Desirable though the above all are, they can often involve programming effort that is disproportionately large relative to the core of the model. One of the valuable functions of a centralized simulation facility is the ability to draw on and reuse past experience in these areas, and thus keep these more cosmetic activities in a reasonable perspective relative to modelling.

Managing simulation

Simulation should be conducted at several levels, and good decisions are more likely to be made concerning the provision of simulation hardware and software if levels in the sense about to be described are explicitly considered. If thinking about this is woolly, then a mismatch between expectations and actualities is almost fore-ordained. Dr Allan Carrie, reporting on a workshop held in 1984 to investigate the expectations of user firms in the UK (Carrie, 1985) summarized the situation nicely:

'It appeared that users were asking for a system that would be extremely powerful, but cost nothing, and could be operated by someone who knew nothing about it!'

Levels of simulation skill

The lowest level is that at which a user with computer access loads a package, and with no more than a general experience of how computer packages work, commences to construct and run simple models. Input is evident from the instructions provided within the package itself, and output is sufficiently clear not to need specialist interpretation. Call this level 1, characterized by minimum study time, 'obvious' function, and diagrammatic input/output, usually strongly related to actual physical layouts of manufacturing objects, i.e. machines, conveyors, elevators and so on. Models at this level cannot be expected to mimic deeply the niceties of the control software, nevertheless they can be valuable in first pass appraisal of a process plan, and in doing 'ball-park' calculations on such matters as line balancing, start factors and broad capacity planning. Level 1 simulation is the level at which sanity checks and balances can be made against simulation studies at higher levels.

Level 2 presupposes more powerful tools than Level 1, and is the level appropriate to more detailed studies of engineering proposals. It is the level at which *engineers* conduct simulations on their own account. Most engineers do not want to become programmers and thereby burden themselves with a further auxiliary skill. Nevertheless they may need to carry out lengthy though intermittent project analysis, and *are* prepared to spend time in either detailed learning of a package which embodies generic models, such as Istel's WITNESS, or in collaborating with a programmer to create tailor-made models in a major simulation language whose construction time and cost is justified by the length of the period over which decision volatility remains in the project.

Level 3 is that of full programming power. It assumes the highest degree of control in making the computer mimic either the real world, or the planner/designer's world, and presupposes sound competence in the underlying simulation language or languages, as well as knowledge and experience in applicable modelling techniques. Level 3 simulation is the level at which *specialists* work. The assumption is that at this level it is possible, in principle at least, to model at any level of subtlety which the available computers allow. Often the tools (i.e. the software packages) used at Levels 2 and 3 will be the same, and the difference in level describes different ways of using them. A far-seeing consultant may also be able to anticipate a model being used beyond the design stage as a tool for assisting line management in production by enabling forward runs into the future to be made starting from current data. This will often presuppose involvement with the model on the part of many people concerned with the operation of the plant or machinery, which in turn means further programming work taking place on 'front ends' that give the model the appearance of a Level 1 tool.

Choosing simulation systems and languages

It is appropriate to consider next the choice of tools from which an organization has to select some for adoption as simulation standards, and to review how such tools have evolved over the years.

Evolution of simulation languages

At first, programs were written in standard languages like FORTRAN and ALGOL, but this soon became very tedious, since the modelling of interactions between e.g. machines and conveyors is a tiresome and finicky business, and is not only prone to error, but has to be repeated over and over again in much the same form for every model. It was logical to construct specialized simulation languages that take care of such standard process interactions once and for all, and require the program writer only to describe in stylized terms the particular characteristics of the system currently being modelled. In the context of such languages the terms 'program' and 'model' become synonymous. The earliest of such languages is GPSS, which dates from about 1960, and which in later versions is still a major force in simulation. This is mainly available on mainframes, although at least one PC version is also available. The early competitors for GPSS were SIMSCRIPT, developed in the USA, and SIMULA, an ALGOL-based language developed in Norway in the 1960s.

In terms of general sophistication, simulation languages lag behind the general run of computing languages. Flow charts, for example, are now almost universally disdained in professional programming circles but are still recommended in one form or another by the promoters of most simulation packages. Again rigorous program specification, a topic that has assumed considerable importance in professional programming, has not yet begun to make any impact in the world of simulation programming. In developing large programs like operating systems, clear distinctions are drawn between coders and testers. By analogy, the development of complex integrated manufacturing systems ought to include complementary teams of designers and simulators.

The graphical techniques recommended by product manuals for model construction help to give insight into the comparative sophistication and capability of the various simulation languages. The writers of early languages such as GPSS recommended that users construct diagrams in which blocks representing stages in the life-cycle of a 'transaction' (typically a manufactured part) were connected to each other by arcs representing sequencing. Flow might or might not take place along the arcs depending on the current state of system congestion. The classical books expounding GPSS are Probst, Bobilier and Kahan (1976) and Schriber (1974). In graphical terms, *Figure 6.4* shows the way a user of a GPSS type of language sees the three-machine model of Section 1.3, with a buffer added ahead of each machine.

Figure 6.4 Three machines in series with buffers

At a later stage of evolution, network queueing languages such as the IBM language RESQ (MacNair and Sauer, 1985), developed by the Research Division at Yorktown Heights, have dealt with the problem of resource contention by using the device of 'tokens', which control the flow in the arcs of a transaction diagram, and overlay it with another system of 'passive' entities. *Figure 6.5* shows how the user of such a language visualizes the three-machine system.

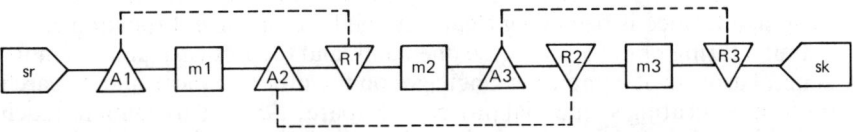

Figure 6.5 Three machines in series with buffers represented by passive queues. A and R stand for 'allocate token' and 'release token' respectively

RESQ tokens have some affinity with the tokens of Petri nets referred to earlier, and it has been suggested that RESQ can provide a straightforward means of implementing Petri net representations. Practical experience suggests that while RESQ is an admirable tool for dealing with systems of moderate complexity, it is not reasonable to try to extend it to the very involved models that arise from incorporating a lot of the fine detail of a complex manufacturing system. This comment would also seem to apply to modelling systems which provide more direct implementations of Petri nets.

Neither GPSS nor RESQ can represent in an immediate way such subtleties as push vs. pull systems which have become all important as JIT (Just In Time), CFM (Continuous Flow Manufacturing), Kanban, and other Japanese-inspired manufacturing philosophies have become subjects of serious study in practical manufacturing. Graphs that describe *processes* and their interactions are the logical development, and these lead to process/entity languages such as SIMPLIX and PERSONAL PROSIM. The latter is described in Sierenberg and de Gans (1986). The nature of the change in viewpoint is illustrated by the way in which the user now employs a single graph as in *Figure 6.6*, which shows a common behaviour shared by the three machines modelled as processes.

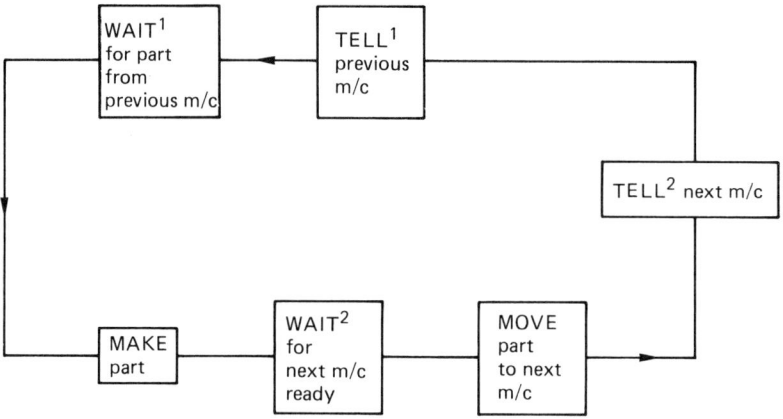

Figure 6.6 Behaviour of machine processes in a push system. Superscripts link each TELL to the corresponding WAIT from which it can release another machine process

This system is a push system. The corresponding pull system is described by the rearrangement of the graph shown in *Figure 6.7*.

The underlying concepts of such process interaction are not new but their significance is becoming clearer in the light of current thinking about manufacturing philosophy on the one hand, and theoretical studies of more general aspects of computing languages on the other, in particular research on communicating sequential processes (Hoare, 1985). This contains much which is of the essence of manufacturing simulation, and suggests that the general-purpose language OCCAM with its emphasis on separating

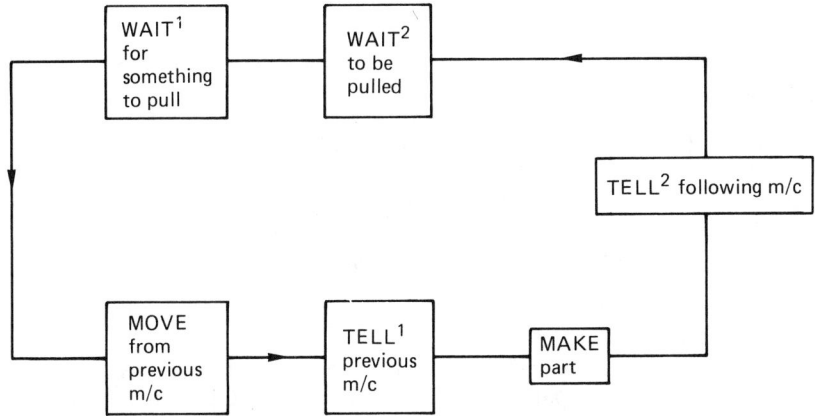

Figure 6.7 Behaviour of machine processes in a pull system

assignment into two stages of input and output along 'channels' may have some relevance in studying the details of key stages in a manufacturing line. This should be especially true of CIM systems where movements occur under direction from control programs rather than from human beings. Increasing automation should be accompanied by increasingly sophisticated computing languages, particularly those used for real-time control. In professional programming, increasing use is being made of specification languages; in manufacturing, specification languages for control systems should be natural candidates for simulation languages.

Available languages
Within an organization, specific choices have to be made of a few languages from among the many different ones now available. These run on a virtually continuous spectrum of computer sizes from large mainframes to small PCs. The machine for which a system is designed, although a highly important consideration in determining what simulation system a particular organization can afford, is much less important from the user's point of view than the classification by level, and then by programming attributes (i.e. transaction/facility, queueing-network, process/entity) within level. Apart from GPSS, most of the mainframe languages have their origins in universities, for example:

SLAM	Purdue
ECSL	Birmingham
SIMAN	Pennsylvania
XCELL	Cornell

To the range of mainframe languages must be added a growing number of PC tools. Less power must naturally be expected from these than from their mainframe counterparts, nevertheless convenience and portability can often make them very useful, either in looking at partial aspects of a design process, or for constructing level 1 type models.

A selection of such systems with their sources is:

PCMODEL	Simulation Software Systems, San Jose
SIMULAP	Fraunhofer Institute, Stuttgart
PC/GPSS	Minuteman Software, Massachussets
WITNESS/SEEWHY	ISTEL, Redditch, England
HOCUS	P.E. Consultants, Egham, England
MAST	CMS Research, Inc., Wisconsin
CAN-Q	Purdue University
PERSONAL PROSIM	Sierenberg & de Gans BV

Both SIMAN and XCELL mentioned above under mainframes also have PC versions.

The various types of languages all have their followings, and the GPSS-like languages in particular have always appealed to engineers more than to programmers. What is important in practice is not partisan devotion to one particular language or system, but rather the quality, skill, and communication abilities of the human consultants involved, together with management dedication to the provision and support of their services.

Examples are given in *Tables 6.3, 6.4* and *6.5* of simplified short programs in the three styles of language. These examples are based on GPSS, RESQ, and SIMPLIX respectively and should be related to the *Figures 6.4, 6.5* and *6.6* respectively.

Generic models
The distinction by programming attribute is not generally of primary importance to the 'client' who sees his input rather in terms of e.g. machine times, buffer sizes, conveyor capacities, labour requirements. Indeed by the time the programmer has completed his work the programming ideas

Table 6.3 Three machine model in the style of GPSS

SIMULATE		
GENERATE	ARR	ARR is the IAT
QUEUE	QUE1	QUE1 keeps statistics of the queue for the facility called MACH1
SEIZE	MACH1	
DEPART	QUE1	Matches the QUEUE block 2 lines before
ADVANCE	TIME1	Service time at first machine
QUEUE	QUE2	
SEIZE	MACH2	2nd machine is demanded before . . .
DEPART	QUE2	
RELEASE	MACH1	. . . first machine is released
ADVANCE	TIME2	
QUEUE	QUE3	
SEIZE	MACH3	
DEPART	QUE3	
RELEASE	MACH2	
ADVANCE	TIME3	
RELEASE	MACH3	
TERMINATE	1	Tally of completed transactions is kept
START	200	Simulation stops when this reaches 200
END		

Table 6.4 Three machine model in the style of RESQ

MODEL: SIM
  NUMERIC PARAMETERS: ARR TIME1 TIME2 TIME3 RUNL

QUEUE: Q1	More properly a 'service centre'
TYPE: FCFS	Short for First Come First Served
SERVICE TIME: TIME1	TIME1 is a parameter (see line 2)
. . . plus similar blocks for Q2 and Q3	
QUEUE: B1	These queues represent the buffers
TYPE: PASSIVE	ahead of each machine
ALLOCATE NODE: GET1	Allocate token(s)
NUMBER OF TOKENS TO ALLOCATE: 1	
RELEASE NODE: FREE1	
. . . plus similar blocks for B2 and B3	
CHAIN: LINE	A chain is a sequence of nodes, along
TYPE: OPEN	with a source generating new units
SOURCE LIST: RAW	of flow into the network.
ARRIVAL TIMES: ARR	
:RAW->GET1->Q1->GET2->FREE1->Q2	Description of *Figure 6.5*
:Q2->GET3->FREE2->Q3->FREE3->SINK	

END

Table 6.5 Three machine model in the style of SIMPLIX

```
SIMPL;
  PROCESS(MACH), PROCESS(ARRIVE), ENTITY(PART);
CONTROL;
DO I=1 TO 3; STARTUP MACH(I);                          END;
  CREATE WLIST(3); CREATE PLIST(3);
  STARTUP ARRIVE;
  TAKE(run-length);                                    ENDC;
BEHAVIOUR(ARRIVE);
  TAKE(iat);
  CREATE PART;
  INSERT PART IN WLIST(1);
  NOTIFY(MACH(1));                                     ENDB;
BEHAVIOUR(MACH);
  LAB: IF not VOID(WLIST(machno)) MOVE FIRST PART
          FROM WLIST(machno) TOPLIST(machno);
                                    HOLD;        GOTO
        ELSE                        LAB;
  TAKE(service time);
  IF machno=3 DESTROY PART;
  ELSE MOVE to WLIST(machno+1); NOTIFY MACH(machno+1);
                                                       ENDB;
SIMPLEND;
```

HOLD and NOTIFY correspond to WAIT and TELL in *Figure 6.6*. PLIST and WLIST are respectively lists of parts in process and parts waiting in the buffers ahead of each machine.

---

should be totally disguised in a user interface that talks in the client's terms. This thinking is carried a stage further in generic model systems. At levels 1 and 2, generic models are the means by which the simulation supplier can provide the kind of user interchange whose expectation has been created by the mass-market computer packages such as word processors and spreadsheets. In the UK the highly popular package WITNESS evolved from the earlier Istel package SEEWHY. SEEWHY is effectively a library of FORTRAN subroutines that the programmer links together to derive a personalized FORTRAN program. The execution of this (i.e. the simulation run) comprises the progressive work-through of a calendar of events, some of which spawn more events to be placed on the calendar for the future, thereby continuing the momentum of the program. WITNESS is a single highly parameterized FORTRAN program whose initial high-level menu identifies the three phases of model construction, viz:

DEFINE
DISPLAY
DETAIL

In the DEFINE phase, the user identifies the major building blocks of the simulation, such as the physical elements listed above, and the control elements, i.e. variables, attributes and distributions. In the DISPLAY phase, the user builds up a schematic representation on the screen of the physical elements established in the DEFINE phase. In the DETAIL phase, the user specifies the timings and routings of the parts as they move through the model. Each element type has its own characteristics, and *Table 6.6* gives a typical menu.

Table 6.6 Machine menu in WITNESS

Name
Quantity
Type
    ★   Single   :
    ★   Batch   : Min/Max
    ★   Production: Quantity/Part type
    ★   Assembly : Quantity
Priority
Labour
    ★   Cycle   : Quantity
    ★   Set-up   : Quantity
    ★   Repair   : Quantity
Input rule
Set-ups
    ★ Basis (Change/Operations/None)
    ★ Time between set-ups
    ★ Set-up time
Cycle time
Breakdowns (Yes/No)
    ★ Time between failures
    ★ Repair time
    ★ Scrap part (Yes/no)
Output rule
Actions (Start/Finish/Breakdown/Repair)
Reporting (All/Individual/None)

The meanings of the individual fields are almost self-obvious. The logic for controlling the simulation is also specified in the DETAIL phase. Following the (possibly iterative) completion of these phases the model is ready either to RUN, that is with full animation, or to BATCH, i.e. execute without animation, which speeds up the simulation, typically by a factor of about 2.

Related techniques

Simulation should not be an activity that exists in isolation from other statistical and operational research type activities, and an account of its role must include some mention of its relationship to these.

Simulation and experimentation
The progress of a simulation can be sub-divided into:

- *Model development*, understanding the situation, program writing and debugging
- *Experimentation*, in which a model proven and seen to be correct by both client and modeller, is run repeatedly with different parameter combinations.

The first stage often has merit in its own right by giving a sharpened understanding of the situation being modelled regardless of whether any explicit results are produced at all. Programming discipline may result in the modeller's insistence in extracting answers to questions which the designer may have been subconsciously avoiding, even to himself. Nevertheless a model which fails to get to the experimentation stage has probably not achieved its full potential. This stage should be undertaken with at least some appreciation of that part of statistics called Design of Experiments (DoE). Usually this means no more than embodying repeated runs within a disciplined framework, or in DoE terminology, conducting factorial experiments.

For example, suppose that COV, IAT and Buffer-size (BSZ) are all variable, and that the aim of a simulation is to study TPUT as the response variable over a range of 5 COVs (0.2,0.4,0.6,0.8,1.0), 6 IATs (5 to 10), and 7 BSZs (0,1,2,4,8,12,20). A total of $5 \times 6 \times 7 = 210$ runs is required to exercise all possible combinations of the parameters, and this is easily realized in programming terms by a set of nested loops:

```
DO COV = 0.2 TO 1.0 BY 0.2;
  DO IAT = 5 TO 10;
    DO BSZ = 0,1,2,4,8,12,20;
      . . .
    END;
  END;
END;
```

where the . . . means whatever triggers a run in the particular simulation system. If the cost of computer time is relatively high, perhaps because the base model has grown to some intricacy, then fractional factorial experiments, or experiments involving highly integrated designs can be

very valuable in extracting the maximum amount of information from the
resource expended at computer run-time.

Continuing the above example, suppose that after a while the ranges of
interest of the parameters have narrowed and that sets of 4 COVs (0.3 to
0.6), 4 IATs (5 to 8) and 4 BSZs (0,1,2,4) cover the range of interest. A
Latin square design could be exploited to organize 16 runs according to the
following pattern:

		BSZ			
		0	1	2	4
	.3	5	6	7	8
COV	.4	6	5	8	7
	.5	7	8	5	6
	.6	8	7	6	5

in which the tabulated entries are the IATs. When the experiment is run,
16 TPUTs are obtained. For definiteness suppose that the values together
with the row, column, IAT, and overall averages are as follows:

		BSZ				
		0	1	2	4	
	.3	1430	1487	1392	1239	1387
COV	.4	1387	1415	1201	1417	1355
	.5	1351	1147	1462	1452	1353
	.6	1012	1211	1481	1456	1290
		1295	1315	1384	1391	1346

Average for IATs   5: 1441   6: 1452   7: 1343   8: 1150

The design of the experiment is such that each of the 12 estimates for the
4 COVs, 4 BSZs and 4 IATs is averaged fairly over all values of the other
two quantities, yet this has been achieved with just 16 runs, as opposed to
the 64 which would be required to examine all combinations of the
parameters.

A word of warning should however be given that the 'fairness' in the
above analysis only applies if there is a linear relationship between TPUT
and the other 3 quantities. The graph of *Figure 6.3* indicates that the
relationships are likely to be far from linear, nevertheless the broad
message about the dependencies on the three quantities being investigated
can be assessed at a first approximation.

Graph theory

Simulation users should be aware of simple graph theory which since the
early 1960s has standardized a number of techniques and algorithms with a
useful place in the repertoire of the simulation specialist. For example,
there is an easily implemented algorithm to determine the shortest path
through a network, each arc of which has a time associated with it.
Consider the network of *Figure 6.8*.

The algorithm consists of establishing a 'potential' of 0 for the source
node S, then at every step establishing the potential of one more node by
evaluating the minimum distance of those nodes which can now be reached

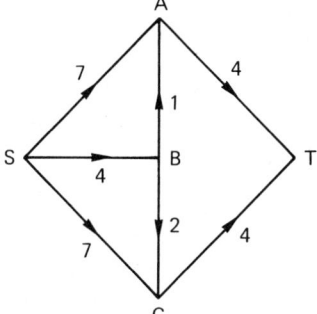

Figure 6.8 Directed network for shortest path algorithm

from points of known potential. For the network of *Figure 6.8* the order in which the potentials of the nodes is evaluated is:

Step	Points directly reachable from:	are:			
1	S	A,B,C	Min(7,4,7)	= 4	so potnl(B) = 4
2	S,B	A,C	Min(5,6)	= 5	so potnl(A) = 5
3	S,B,A	C,T	Min(6,9)	= 6	so potnl(C) = 6
4	S,B,A,C	T			so potnl(T) = 9

Suppose now that a factory contains 40 machines each of which is dedicated to running just one of 12 processes. The parts require to have the 12 processes applied in sequence, so for any particular process, just one of the set of machines carrying out that process must be visited. The factory management may place the 40 machines anywhere on the factory floor, and the problem is to find a deployment that minimizes the overall handling time. This is a straightforward extension of the small example given above, and requires only an expansion of the data, with no change in the algorithm itself.

Linear programming
Another category of techniques comes from the area of mathematical and linear programming, where the dominating idea is optimization subject to constraint. An example concerns the determination of the optimum pattern of vehicle arrivals to achieve a projected target number of pallets unloaded daily at a factory. Pallets arrive in a variety of vehicle types whose capacities range from 1 to 40 pallets. The time taken to unload a vehicle is proportional to the number of pallets it carries, and the number of unloading bays imposes a limit on the number of vehicles that can unload simultaneously. The road space available for vehicles waiting to unload is also limited. To meet the production objectives of the factory, particularly those associated with JIT, there are minimum requirements for the numbers of pallets arriving within each of a number of time periods within the working day. Because the larger vehicles come from very long distances, it is less easy to constrain their arrival times, and it must be assumed that some proportion of this traffic will arrive at times outside the control of the factory management.

The conditions spelt out in the last paragraph can be translated algebraically into *constraints* within which a range of solutions has to be sought that achieve the targeted arrival pattern. The algorithms which carry out this sort of analysis were initially developed towards the end of World War II and are now well understood, and implemented in packages such as MPSX (Mathematical Programming System Extended) which is IBM's standard LP offering, and SCICONIC which is a similar package from Scicon of Milton Keynes, England. Useful models in the area of CIM are typically very small compared to the general run of linear programming models, and run times correspondingly short. They can thus be very valuable for homing in on a range of admissible solutions, where initial specifications are fairly vague. Then a simulation model can take over, and examine in finer detail the consequences of setting parameters to a more limited range of values. A book that gives an excellent survey of the modelling possibilities afforded by linear programming is Williams (1985).

Manufacturing laws
An as yet relatively untapped area in which simulation can contribute to manufacturing science is the evolution of Manufacturing Laws akin to Newton's Laws in physical science. A few already exist, e.g. Little's Law, but many more remain to be discovered. Little's Law is conveniently remembered by the mnemonic

$$L = IT$$

where L stands for the length of a queue, T the average time spent queueing in it, and I is the average inter-arrival *rate* of objects which come to the service centre. One possible way to exercise simulation in this role is to use it to construct tables (perhaps embodied in computer programs) that the design engineer can consult to obtain quick answers to questions concerning how much resource has to be supplied to counteract an anticipated degree of variability.

Making simulation effective

The problem of making simulation effective is much broader than that of choosing a few simulation packages and allocating some people to the task. It must be the subject of ongoing management attention and involvement, and conscious discipline must be imposed to achieve any worthwhile results. The concluding section of this chapter highlights some of the critical management considerations.

Education for simulation
The purchase of a simulation system without provision for education and training is simply a waste of money. At levels 1 and 2, there are likely to be quite large numbers of candidates requiring fairly short bursts of instruction, and an on-going in-house programme of short courses is a reasonable prospect for even relatively small organizations. At level 3 however, the numbers are much smaller and the education of simulation specialists is necessarily much more of an ad hoc affair which is dependent on the choice of simulation tools made by the organization, and may perhaps involve purchasing education from outside.

The importance of timing

No matter how great the simulation skills available, they cannot be used to good purpose unless they are brought to bear at the right time, that is neither too late after designs have become inflexible and unable to respond to changes suggested by simulation, nor too early when the design is too volatile for a model to stabilize. Thus manpower planning is a crucial part of planning for simulation. Clearly there must be a flexible pool of skilled people with the combined skills of modelling, programming and statistics, deployable at short notice. This is difficult enough for large organizations; smaller ones may have to rely on buying the service of consultants.

Cost justification

Simulation provision planned as above makes it difficult to achieve accurate costings of the worth of simulation in terms of savings achieved. If interaction between simulators and designers is high, then simulation is likely to be just one of several factors influencing changes in design, and consequently it may be difficult to separate out its contribution to cost savings. Sometimes, of course, simulation may be carried out with an explicit objective of investigating the costs of alternative designs, in which case the cost benefit may be easier to establish.

The value of total involvement

Simulation is often most effective when the area of involvement with a model or models spreads beyond the immediate boundaries of consultants and design engineers. A good model should not come to the end of its usefulness when the design cycle is complete. Ideally there will be a 'follow-through' phase as suggested earlier, in which the model is 'front-ended' and installed on a widely-available computer system so that people involved with production and line management can run it under hypothetical conditions, preferably from starting points of real time data. In this way simulation becomes intimately involved with day-to-day control, and its benefits can be long lasting. It is hard to overstate the effect that decisions made by human managers have on the actual running of even the most highly automated lines. Until the computer is *really* in complete control for extended periods (the ultimate if perhaps distant goal of CIM) it will be impossible to judge after the event whether the predictions of simulation were in fact borne out. However the cooperation of the people actually operating the system is invaluable in establishing the creditability of simulation. It can also help overcome the 'Cassandra effect' experienced by all forecasters: that is, if the results of simulation match engineering expectations, it will be judged a waste of time and effort, if they do not, it is suspected and therefore ignored.

Choosing points to attack

There is a danger in indiscriminate simulation with too general a brief given to the simulation specialist. With a large project it will frequently be the case that substantial areas of the plan do not justify simulation, and thought given on what to *exclude* as well as what to *include* within the scope of simulation can often result in limited resource being directed to better effect. It is a common mistake to try initially to construct a massive model with a one-to-one representation of an entire production area. In general it is much better to work instead for a full understanding of critical areas, particularly those which are possible bottlenecks, and where the distinctions between e.g. push and pull philosophy can produce appreciable performance variation. Whereas 'islands of automation' is on the whole a derogatory term in CIM, 'islands of simulation' should be regarded as a desirable goal.

Scheduling, simulation and control

It was argued above that a linear programming package is a natural complement for a simulation system at the *planning* level. At the *operational* level, LP has long been recognized as a tool for scheduling job-shop type operations, and simulation has a natural though less frequently implemented role of predicting the consequences of a particular shop-floor schedule and evaluating scheduling alternatives.

In many simulations the specialist comes to a point where in mimicking the real-world system, he or she is retreading (or possibly pretreading) the path of the writers of the underlying control software. There are two ways in which this can be exploited, one as a check on the effectiveness and quality of the control software writer. If the work of the latter results in performance seriously below that predicted by simulation then either there is a deficiency in the modelling, or the control software is not extracting the best from the machinery it is controlling. On the other hand if the simulation is carried out in advance of the control software, then it is possible that the algorithms of the former can be exploited in the writing of the latter.

In taking a broader view of a CIM operation, some software vendors market sets of complementary packages, which enable a measure of consistency to be achieved in the three areas of scheduling, simulation and control. This can be particularly valuable where data is to be shared, and all three activities use the same databases. Integration of this sort can be extended further to include inventory control systems, material resource planning systems, plant logistic systems and more recently knowledge-based systems such as KEE, Knowledge Craft and ART.

The theme of this book is integration, and the aim of this chapter has been to highlight the role and methods of both the simulation specialist and the simulation system, and the ways in which the former should be integrated into the engineering team which plans and runs a CIM installation, and the latter into the software which operates and controls it.

Further reading

ABRAHAM, C. T. and SCIOMACHEN, A. (1986) 'Planning for Automatic Guided Vehicle Systems by Petri Nets', *IBM Yorktown Research Report no. RC 12288*

BOBILIER, P. A., PROBST, A. and KAHAN, B. C. (1976) *Simulation with GPSS and GPSSV*, Englewood Cliffs, NJ: Prentice-Hall

CARRIE, A. S. (1985) *FMS simulation: needs, experience, facilities*, Proc. 1st Conf. on Simulation in Manufacturing, Stratford-on-Avon, UK

CONWAY, R., MAXWELL, W. L. and WORONA, S. L. (1986) *User's Guide to XCELL factory modelling system*, Redwood City, CA: Scientific Press

DANGELMAIER, W. and BACHERS, R. (1985) *SIMULAP — a simulation system for material flow and warehouse design*, Proc. 1st Int. Conf. on Simulation in Manufacturing, Stratford-on-Avon, UK

HOARE, C. A. R. (1985) *Communicating Sequential Processes*, Englewood Cliffs, NJ: Prentice-Hall

KENDALL, D. G. (1953) 'Stochastic Processes occuring in the Theory of Queues', *Ann. Math. Stat., 24*, pp 338–354

MacNAIR, E. A. and SAUER, C. H. (1985) *Elements of Practical Performance Modelling*, Englewood Cliffs, NJ: Prentice-Hall

PAGE, E. (1972) *Queueing Theory in OR*, London: Butterworths

SCHRIBER, T. J. (1974) *Simulation using GPSS III*, New York: Wiley

SIERENBERG, R. W. and DE GANS, O. B. (1986) *Personal PROSIM Textbook and User Manual*, Waddinxveen, Netherlands: Sierenberg and de Gans

WILLIAMS, H. P. (1985) *Model Building in Mathematical Programming*, 2nd edn., New York: Wiley

It will be different with CIM

Chapter 7

Components of a CIM architecture

What is a CIM architecture?

A CIM architecture is a framework, providing a structure of computer hardware and software, with the appropriate interfaces, for computer systems in manufacturing companies. The architecture should be valid for not less than ten years, i.e. probably to the end of the century. Some of the interfaces will be provided by international standards and by *de facto* industry standards. Others will be determined directly by the suppliers of hardware and software. The business problems of manufacturing industry and the developments in technology will be the major driving forces. There are many different aspects to the subject; this chapter will try to take the common ground and define the principles of a general approach.

A good starting point for an architecture is on the shop floor, since there is almost universal agreement that a LAN (Local Area Network, briefly introduced in Chapter 4) is required there. LANs are conventionally used in the office environment; there is a reasonable consensus that in most manufacturing environments the LAN will need extra function, called therefore an ILAN (Industrial LAN — the suitability of different types of LAN and ILAN will be discussed below). Into this ILAN will be plugged shop-floor computers, including for instance workstations and controllers of FMS (Flexible Manufacturing System) cells. A link will be established to mainframe computers, which will in general perform business planning functions. This is illustrated in *Figure 7.1*. The mainframe computer could plug directly into the ILAN, or be attached via a terminal.

Although the use of an ILAN is almost universally accepted, there is less consensus on the number of 'levels' of hardware which should be shown in *Figure 7.1*. Effectively there are three in the figure: mainframes, cell controllers and device controllers. It is not uncommon to see such diagrams showing five levels (say by adding a 'corporate' level at the top of *Figure 7.1*, and an 'area' level between the cells and the ILAN). However, the number of levels should not become a major issue, if only because it is far from clear what actually differentiates one level from another. Since much of the data needs to be common to all levels, in an integrated system the user would usually want to see only one level.

'Plant logistics' is shown running on a computer connected to the ILAN; these planning and control functions could of course be carried out on the

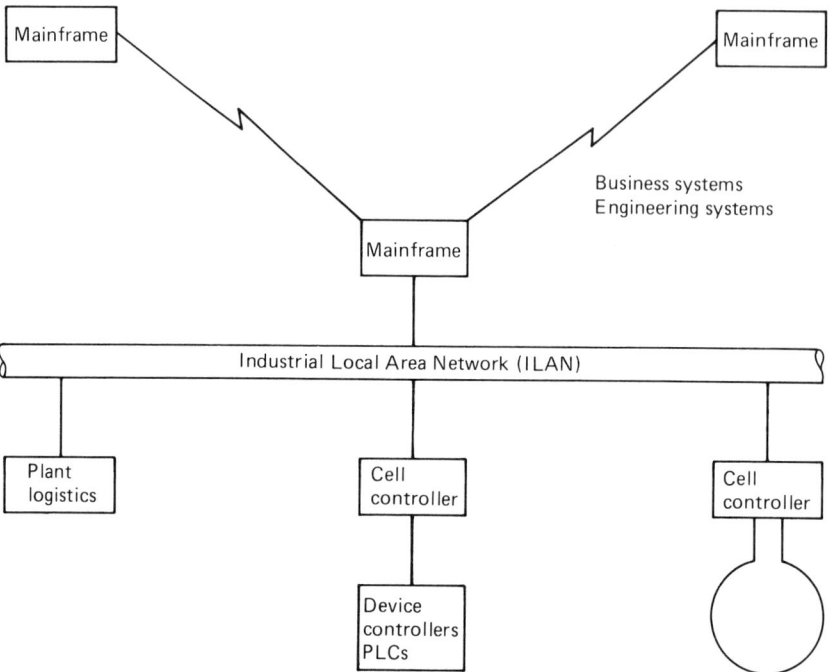

Figure 7.1 CIM hardware hierarchy

mainframe. Flexibility to move this function to different computers would be important.

Figure 7.1 also shows mainframe computers connected together. This will happen when different sites are to be connected (WANs, Wide Area Networks) or for instance if the CAD system is on a different computer from the business planning functions. (As discussed in Chapter 4, two or more computers from the same manufacturer linked together, would be regarded as one island of computerization in this context — clearly the manufacturer provides the communications technology). There are therefore two points in *Figure 7.1* that require hardware of different manufacture to interface:

• between mainframe computers,
• into the ILAN on the shop floor.

Although there are many proprietary means of making these links, the major mechanism for connecting different types of computers is by using international standards (discussed in some detail below).

Software is, of course, also an important integration tool. Common operating systems are being developed for different computers. Decision support systems will include the use of databases to integrate existing systems. Expert systems and artificial intelligence seem to be essential to analyze the vast amounts of data that will be available from these linked systems. Standards for data exchange between different CAD and graphics systems are also being developed, and the integration functions of workstations need to be considered. These are all large subjects, and will be explored further in this chapter as components of a CIM architecture. There must be some hesitation in adding to the enormous number of flow charts, cylinders and pyramids which have been so assiduously produced to graphically represent CIM, so the figures in this chapter will be kept simple. However, before getting into the detail, it is important to discuss why an architecture is necessary at all.

Why have a CIM architecture?

The most important single point about CIM architecture is that it should be seen as a business tool, not as an end in itself. This architecture is not going to be perfect or complete. Just as domestic power standards vary around the world and there are many different versions of BASIC, so there will need to be compromises in CIM architecture, because of the cost of restructuring existing systems, the uniqueness of each manufacturing company and the competitive market place for computer-based products.

Even with these limitations, an architecture should have considerable advantages in enabling a CIM system to be adapted more easily to new business opportunities and to changing technology. If there is a quality problem, more data should be easily collected and analyzed. If another business is taken over, immediate computer-to-computer communication should be possible. If workload in the factory increases, a new package to plan capacity in more detail should just be 'plugged in' (the standardized application interfaces implied by this last example do not yet exist, but may well evolve with the increasing use of relational database and query systems).

A proper structure (i.e. an architecture) is essential to effective management and use of computer systems. An architecture implies the opportunity to use standard solutions, usually by installing software packages (rather than by writing one-off systems). Apart from often having direct cost benefits, software packages have the advantage that external education for the users and the IS staff will be available. When additional software is written in-house, it can take advantage of the standard interfaces.

An architecture requires a structured plan and a planning process, so that systems can evolve. In an environment of increasing user expectations and rapidly changing business problems, an architecture will therefore be a mechanism of change for any users dissatisfied with the limitations of an existing system. If the change required is simple, the architecture should allow it to be quickly implemented.

Conversely, without an architecture, computer systems will be much restricted in flexibility. New hardware or software which has potential value (and which competition may be using) could appear uneconomic because of the conversion and modification work necessary; thus, without an architecture, systems become progressively more out of date.

There are, however, factors against an architecture. Some changes may take a long time simply because they have to be checked against the architecture and future strategy. Also the architecture may be too complex for some business needs. For instance the architecture might assume a multi-level bill of material with extensive engineering change facilities —

necessary to many companies — but maybe more trouble than it is worth if the company makes drinks cans which have two components. But what if other products are introduced, and isn't the multi-level bill of material needed to hold the structure of the machinery on the manufacturing lines for the maintenance engineers?

Another possible argument against having a CIM architecture suggests that innovation is a step function, and that the best business strategy is to replace a computer system outright (like an old car) and to buy a completely new one. However, people cannot be retrained overnight, and this strategy will not work if the CIM system is so extensive that replacing it all at one time just can't be done. When parts of the system are to be replaced, an architecture is needed to ensure the new fits with the old. Furthermore, some costs, for instance installation of cabling, should only be incurred once.

Hence, an architecture is essential for a CIM system to be effective, and therefore for a manufacturing company to survive. The components of a CIM architecture will be described in this chapter, and developed in the following chapters. As mentioned above, CIM architecture requires the continual development of international standards.

International standards and OSI

International standards are needed if computer systems from different manufacturers (as in *Figure 7.1* and discussed above) are to be connected together.

Development of international standards

These standards are agreed formally by organizations that are independent of competing manufacturers and competing nations. Hence the ISO (International Organization for Standardization) works through committees of representatives of the national standards organizations of each participating country. These in turn (ANSI in the USA, BSI in the UK, DIN in Germany) are formed from representative industrial organizations within the country, and not from individual companies.

There are real difficulties in establishing international standards for leading-edge technologies. As in many specialist subjects, there is a limited number of experts; the experts have 'real' jobs to do and can only spend part of their time working on standards committees. Hence the committees only meet occasionally (say, every two to three months) and the elapsed time for the process of defining, discussing, modifying and agreeing international standards can usually be counted in years.

However, this may be time well spent, since loosely-defined standards can cause considerable problems. If there are too many options (or any ambiguities) within the standard, then it will be possible for two different products to each meet the standard and yet not be able to communicate with each other.

Physically, a standard is a document giving detailed specifications of hardware and software protocols (e.g. the fifth bit is to be set to one if the data is longer than 128 bytes). ISO standards usually begin as working papers that are published as Draft Proposals (DP) when agreed by the appropriate ISO working group. These are then reviewed and may be extensively changed by the ISO member countries, after which stage they become Draft International Standards (DIS) and are relatively stable. Finally, the DIS becomes a full International Standard.

Testing products against a communications standard must include conformance (i.e. does a product conform to the standard?) and interoperability (i.e. do different products that conform to the standard actually work with each other?). Further testing is then needed to identify any performance, security or reliability problems in the products that would inhibit successful use of the standard.

The traditional function of the standards organizations to harmonize and approve established working practices is being very much stretched in these areas of new technology. Given the speed of technical development in CIM, it is possible that a standard could be obsolete by the time it has been formally accepted. The standard could be equally unusable, if, in an effort to speed up the definition process, a courageous but wrong judgment were made about future developments, and a standard were based on what proved to be inadequate technology. As a result of such considerations, the standards organizations, assisted by user organizations and pressure groups, are implementing speedier ways to produce standards.

In practice, many hardware and software products have become *de facto* standards by virtue of their wide use and acceptance. For such products, the interfaces are often published and other suppliers are able to provide complementary (and competitive) products. Some products develop, or are extended, into international standards.

Many suppliers of computer systems have their own proprietary interface protocols, which may well not be adapted to international standards because of cost and the fear of loss of control. From the point of view of CIM architecture such systems, unique to one manufacturer, will be considered as 'islands of computerization'. The interest will be on the standards now emerging to integrate these different proprietary systems. The most important of these is the OSI reference model.

Open systems interconnection (OSI)

The use of the letters OSI for a major set of ISO standards seems an interesting choice; the two acronyms certainly provide a trap for the unwary. The OSI reference model, on which work started in 1977, provides a seven layer communications architecture that is shown in outline in *Figure 7.2*.

The seven-layer hierarchy divides up the complex functions needed to interface many different computer systems for different applications. Separate functions such as controlling the physical connection or providing an interface for user programs, being in different layers (layers 1 and 7 respectively), can be performed by different hardware and software. This is useful in allowing future technical developments to be incorporated. If (and when) fibre optic technology, for example, is introduced as the physical medium (layer 1), most of the layers above can run unchanged. A

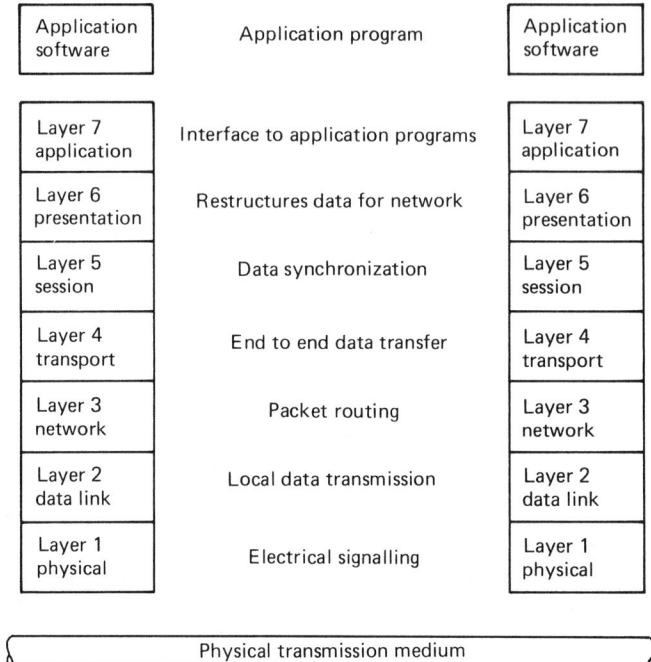

Figure 7.2 The OSI reference model

common analogy is that telephone numbers do not have to be changed when the call is transmitted by satellite rather than conventional land line. However, introducing the extra layers creates an extra, very large, interfacing and software testing problem.

When the OSI reference model, and products based on it, are fully tested and established, a manager or professional learning about CIM architecture will hopefully not need to understand the functions of the different layers. However, until that time (which is some years away) decisions about computers in manufacturing industry are likely to involve some understanding of the different layers (although some people may wish to skip the detail at first reading).

Layers 1–4 transmit the data from system to system without error, whilst layers 5–7 interpret the data stream and present it to the user. Within these layers, OSI provides a selection of standards that can be used for different communication purposes. Two types of standard are associated with each layer of the reference model. The 'service specification' defines the facilities and functions offered to the layer above (which, if a full seven-layer OSI system were supplied, would be transparent to the user). The 'protocol specification' of a layer defines the rules and conventions for communicating with the corresponding layer on another machine.

In broad overview of the seven layers:

● *Application layer* is the interface between OSI and the applications that use it. Standards are concerned with data exchanges and the services provided: the common elements for the management of the OSI resources (CASE, Common Application Service Elements); remote file manipulation (FTAM, File Transfer Access and Management); distributed processing (JTM, Job Transfer and Manipulation); message transfer (MHS, Message Handling System, supporting the CCITT X.400 recommendations); manufacturing messages (MMS, Manufacturing Message Specification).

- *Presentation layer* selects the syntax to be used for data transfer. Syntax is precisely defined within OSI as a special language (Abstract Syntax Notation One, ASN.1) that ensures common representation between different manufacturers. Data is restructured to/from the format used in the network.
- *Session layer* provides facilities to manage and co-ordinate the dialogue between application processes, including restarting a dialogue if there has been a temporary communication failure. The expedited data transfer service allows a small amount of data to receive top-priority transmission.
- *Transport layer* provides reliable transfer of bit streams between end systems. Five 'classes' of protocol are defined depending on the quality of service required by the upper layers.
- *Network layer* sets up the communication path between nodes. In connection-oriented services, the connection is established for the duration of the communication, as opposed to connectionless operation, where no prior connection is established. The CCITT X.25 recommendation supports connection-oriented service, typically used for wide area networking (WANs).
- *Data link layer* transmits the bit streams through the physical medium. It is divided into two sub-layers, logical link control and medium access control.
- *Physical layer* activates the mechanical and electrical operation of the network. Standards include ISO 8802/3 (CSMA/CD), 8802/4 (Token Bus), 8802/5 (Token Ring) which are discussed further below.

What happens physically to the user data from the application program is that extra data (called header data) is added to it on the way down the layers and stripped from it on the way up.

Functional standards
The selection of standards within the different layers can be changed by the relevant ISO committee, i.e. new standards could be added to the OSI reference model, increasing the choice of standards available.

For a specific function or application, a set of standards would be chosen, called a profile. This has been done for MAP (Manufacturing Automation Protocol), for example. Since many of the standards in the OSI 'stack' contain options, just specifying the standard is not sufficient. Inter-operability tests between two products could fail because suppliers have exercised their right to select different options from a standard. Functional standardization is the process of selecting a profile (or profiles) of standards and options to form a practical, useful system.

Local area networks

Figure 7.1 shows LANs, specifically ILANs, as an essential component of CIM architecture. LANs provide peer-to-peer (i.e. any to any) communications between attached stations. In the context of OSI (and the MAP initiative by General Motors, of which more below) they have only recently, in the mid-1980s, become a significant factor on the manufacturing shop floor.

Advantages of LANs

LANs provide a common interface that allows suitably configured terminals and computers to be easily plugged in and disconnected (in a fashion roughly comparable to a domestic electrical system). Since data is sent from one point on the network (called a node) direct to another node, the switching and load balancing function of the central computer is avoided. Operating system type software is, however, still required for networks and is discussed below. A LAN should usually be installed only once in a building, thus avoiding the costs of rewiring.

If a LAN system does not conform to an international standard, there is an implication that connecting other equipment and other networks could be difficult, i.e. expensive and/or time consuming. This will be important in a factory, since few people can be certain what manufacturing equipment and data collection might be needed three to five years in the future. It would be highly undesirable to be limited in choice of shop-floor equipment by the shop-floor LAN.

LAN technology gives the opportunity for resource sharing of printers and discs, so that printing costs can be reduced and common files maintained; a 'file server' is a computer on a network dedicated to providing disk storage for other nodes of the network.

High reliability and serviceability are provided by the flexibility of LANs. 'Vertical' communications to mainframe computers and 'horizontal' communications to other networks are provided by gateways, routers and bridges. LAN cables are capable of very high data rates (e.g. nominally 10 Mbits per second for a broadband MAP network, 100 Mbits per second for fibre optic cables), which is important for transfer of large files of production data and CAD applications.

Application program design
The opportunity to send data direct from one node to another introduces a different approach to the design of applications. Conventional application design would send all the data to one controlling node on the network, where the data would be processed and instructions subsequently issued to cells, robots and machine tools. However, to fully use the network capability, each node would send data only to those nodes which use that data, and receive data only from the relevant nodes. For example, when a machine tool finishes machining a part, instead of sending a signal to a controlling node which in turn tells the robot to unload it, the machine tool sends a signal directly to the robot. The robot is not of course likely to move until it has received other data, perhaps from limit switches via a

field bus, that the conveyor is clear to receive the completed part. In practice it is doubtful if many installed networks have yet taken advantage of this opportunity.

New program design and testing techniques are introduced by this approach. It may be necessary for back-up and recovery purposes to send duplicates of all messages to a 'back-up' node which will hold a database of status information. Collection and analysis of statistical data for process quality purposes would be a parallel application.

LAN topology and protocols

Topology describes the physical layout of the network cabling. Protocols are the procedures and signals used by the different terminals on the network to communicate with each other (there is a comparison with use of the word protocol in the diplomatic context to enable communication between rather disparate parties).

The basic mechanics of the connection from the terminal to the network will be a PCB (printed circuit board) plugged into the back of the PC or other computer, with additional software. The balance between hardware and software (and indeed whether the PCB is replaced by a chip) will depend on the state of technology and the degree of confidence in the stability of the protocols. A standardized plug interface will provide connection to whatever cabling is used.

The mainframe-based communications systems that were the basis of the manufacturing information flow described in Chapter 5 had a central point of control based on a 'star' type of configuration. All communication was from or to the mainframe, which was always in charge. In a peer network, physical communication can be provided by a ring or a bus. A ring is a loop of cable, around which the data is transmitted. A bus, as described in Chapter 4, is a single run of cable to serve as the network backbone, with additional hardware and software to send data up and down the cable.

For both the ring and the bus, there needs to be a mechanism of control, i.e. the protocol or media access method. One approach is to define a token, with some conceptual similarity to the baton in a relay race. The node holding the token (which exists as a bit pattern in the memory of that node) has control of the network. It can attach data to the token and send it to another node. This protocol on a bus is called Token Passing Bus and is described in ISO Standard 8802/4 (also IEEE 802.4, the similarity of the numbers indicating that the ISO adopted the IEEE standard), which is in Layers 1 and 2 of the OSI Model, as shown in *Figure 7.3*. The Token Ring standard is ISO 8802/5 (IEEE 802.5). It is the basis of many office communication systems, and is often used on the manufacturing shop floor in clean environments.

In the ISO standard 8802/3 (IEEE 802.3) nodes send data as and when they wish. If there is a collision (i.e. if two different nodes transmit simultaneously) then each node retries transmission after a random time interval. This access control protocol is described by the acronym CSMA/CD (Collision Sense Multiple Access/Collision Detect). It is used in Ethernet[TM] and in TOP (Technical and Office Protocol, see below).

A comparison can be made between the Token Passing and the

CSMA/CD protocols. The Token Passing Bus (8802/4) is 'deterministic', which means that once a node decides to send a message, it is possible to calculate the elapsed time before the message can be transmitted. This can be very important in certain shop floor situations. CSMA/CD is not deterministic since it is not possible to calculate how often a message will have to be retried (because of collisions with other messages). The time for any retries of transmissions is a function of the load on the line and cannot be determined in advance. Furthermore, at about 30–35% loading, by conventional statistical calculations, response time in a CSMA/CD system can start to increase quite sharply, as the nodes devote more and more time to failed starts.

Keeping the load below about 35% of capacity may of course be manageable; a classic need for a deterministic system, the requirement for NC machine tools to receive the next block of data within a fixed (and very short) time period, can be obviated by putting additional cheap RAM memory on the NC controller. However, vital messages ('There is a fire risk if this valve is not adjusted immediately') must be transmitted. Hence, the case for a deterministic system is a major factor in choice of shop-floor LAN, but perhaps not overwhelming.

Relays are used to transfer data between types of LAN; gateways, routers and bridges are OSI device types. A gateway is a 'dual architecture' device, which by containing two architectures is able to interconnect two different networks. Bridges connect two identical ISO 8802 networks. For networks with different physical media and access methods (i.e. OSI layers 1 and 2) to be connected, a router would be used. LANs can be directly linked over telecommunication lines, using say X.25 protocols.

Office LANs
Office networks are well established and there is a multiplicity of products available, some based on international standards and others on proprietary protocols. The selection of the office LAN may not be an important issue for CIM unless there is an implication that the office network will be extended to the shop floor.

TOP (Technical and Office Protocol) is an OSI implementation for office communications which is closely associated with MAP (e.g. joint MAP/TOP user groups). However, it is not specific to manufacturing industry.

MAP (Manufacturing Automation Protocol)

Work on MAP was initiated in 1980 by General Motors (as a user of computers, not a computer manufacturer), because they were able to anticipate growing problems of linking shop-floor computers unless there were some form of standardization. At this time, GM already had 20000 programmable controllers, 2000 robots and over 40000 intelligent devices in its factories around the world. By 1986 the MAP user groups around the world (USA, Europe, Japan and Canada) involved most other manufacturing companies and almost all suppliers of shop-floor computing.

MAP is an implementation of the OSI model, which is to say it is a set of standards selected from the OSI model. It has been demonstrated with a wide range of equipment at a succession of major exhibitions. It has been through several releases, stabilizing for a while on Version 2.1 in 1986, with MAP 3.0 available in 1988/9. These levels are important because MAP products made to the different levels use a different selection of standards (reflecting the fact that the standards themselves are developing). The timing of the availability of the different levels will affect decisions on implementation of MAP systems.

In addition to the selection of standards from the OSI options, MAP has required the definition of some functions not currently available within OSI. This was done by the MAP user groups and suppliers, enabling work to proceed more quickly than if new international standards had to be defined.

MAP has been criticized for the relatively high cost of installing broadband cable and the connections to it, and because the full seven layers of the OSI Model may make data transfer too slow in certain time-critical situations. The carrierband option of MAP is a partial answer to the first criticism, though it sacrifices the multi-channel capability of the broadband cable. MAP costs will fall with high-volume usage, and in any case can be held to be low if, rather than just comparing hardware connection cost, the total system cost including software is considered. This is based on some of the functions in MAP, particularly MMS (Manufacturing Message Specification) which assists programming in a manufacturing environment.

MMS is an important part of MAP. It is developed by an EIA committee and defines protocols for network services such as device status and control, program load, event management, alarms, job queueing and journal management. An ISO notation called ASN.1 is used to define MMS because it allows such functions to be described independently of any computer language; a programming interface then has to be written for each computer language which supports MMS. The functions of MMS will affect application design and will reduce the application programming required to instal MAP systems. MMS may well become more generic and extend to applications beyond the factory floor.

For time-critical data transfer two extensions to MAP have been developed, MAP/EPA and MiniMAP. A MAP/EPA node (Enhanced Performance Architecture) has a full MAP version on one side (as in *Figure 7.2*) and a 'collapsed architecture' version on the other side using only Layers 1, 2 and 7. This allows communication on one side with full MAP nodes and gives rapid data transfer on the other side. MiniMAP has the same collapsed architecture as MAP/EPA only, so it is not OSI compatible; it can communicate with OSI systems via MAP/EPA. Although the three-layer approach provides an immediate solution, the full seven-layer OSI implementation of MAP is likely to become quicker and cheaper with improvements in hardware (and software). Whether the three-layer approach will be sufficiently established before the perform-ance of the seven-layer version catches up is very difficult to forecast confidently, a nice example of the uncertainties in this subject.

The question of which physical medium (i.e. cable) to use for a shop-floor ILAN, implicit to this discussion about MAP (and also relevant to office LANs), needs further consideration.

Choice of cable for the shop floor
Broadband coaxial cable is used in MAP and has a number of advantages in the manufacturing and process industries. Installation and maintenance of broadband is well understood, since it is commonly used for cable TV (CATV, Community Antenna TV). The word 'broadband' signifies that the coaxial cable can carry a broad frequency band. It has a high capacity, defined as a frequency range, of 50–450 MHz (millions of cycles per second). Frequency multiplexing transmission enables multiple channels to be on the cable at the same time (e.g. both Ethernet and MAP). Voice channels, video pictures for security systems and the volumes of data required for CAD and vision data could be accommodated. Only one cable is required, since it is bus topology (see above) and coaxial cable is safe from electrical interference. Broadband is suited to a LAN covering the typical area of a factory (i.e. a few miles).

Twisted wire pair cable is widely used for computer communications in the office, where there is no electrical interference, and can be suitable in clean manufacturing environments. It lacks, however, the capability to carry other signals.

Carrierband transmission can be on ordinary coaxial cable of the kind used in television aerials; it is a low-cost option (no head-end remodulator for instance) but is single channel, i.e. the cable cannot be shared.

Fibre optic cables are widely used in telecommunications transmission of high-volume data, and are being introduced onto the factory floor (and into MAP 3.0, as an appendix). An optical fibre is a rod of silica glass (finer than a human hair) along which infrared light travels. Fibre optic cables are free from electromagnetic interference and very secure (i.e. against tapping). The disadvantages are cost, which is rapidly falling with increased production, and the difficulty of making connections to a fibre optic cable. Simpler connection techniques are being developed, however, and the use of fibre optics on the factory floor is likely to increase.

The total requirement for data transmission within a business leads to what is sometimes called the 'intelligent building', whereby cabling for all anticipated future requirements, including computing, telecommunications, heating controls etc., is installed as part of the fabric of the building. This puts cable costs in context as part of a long-term investment.

Workstations

As described at the beginning of this chapter, workstations can be attached to a LAN or direct to a mainframe. A workstation is usually a single user terminal with local processing power; it can be a significant component of CIM architecture. A unique definition of a workstation is not straight forward, since the word also appears to be a generic term describing the current top of the range PCs. A workstation will have enough computer power to do considerable 'local' processing (i.e. computer programs running in the workstation, not directly dependent on a central computer). It is generally distinct from a host-dependent graphics terminal (with processing and database access performed only on the host), though of course this distinction may not matter much to the user, and networked workstations may be difficult to distinguish from graphics terminals in some applications.

Workstations can have specific integration functions, with interfaces to link to different network protocols and host computers, often simultaneously. Windowing techniques are used to handle multiple sessions; windowing standards are emerging that will radically affect the development of application software.

Workstation interfaces (hardware and software) can be provided as add-ons to the original workstation. It is worth checking which interfaces can operate simultaneously. Some may require the user to load different system software to run different sessions.

Rating the power of a workstation is not straightforward. MIP rates in the three-figure range will become common, although MIP rating does not necessarily relate to performance in real work (as discussed in Chapter 4). Another rating is the capability to draw vectors or polygons on the display, the power required to rotate wire frame or solid images being of course supplied by graphics chips in the workstation, not by the host.

The display screen will usually have a high pixel resolution (say 750 × 1000 or above, preferably in proportion to the dimensions of the display screen). This high function implies a large memory (measured in megabytes), large fixed storage (measured in hundreds of megabytes) and a 32-bit processor. The operating system may well be Unix based (see below). A full range of input devices (keyboard, WIMP etc, see Chapter 4) is usually provided. The system could benefit from being microprogrammable so that time-critical applications can be optimized. High-performance floating-point calculations will be supported.

General Motors has initiated an Engineering Workstation Platform Specification Project (EWPSP) that provides guidelines for workstations in the form of a list of features categorized as 'required' or 'desirable'. This is a further example of the need for some sort of agreed 'standard' (to which vendors can work and other users thankfully copy) in less time than the necessarily long process of consultation for formal international standards.

Software is important to the effective operation of a workstation. Displaying the output from separate programs on a workstation using windows is easier than the transferring data between these programs. The return of data to the host, after manipulation on the workstation, can require complex software on both workstation and host. Software as an integration tool needs to be considered.

Software integration

Business integration for the people working in manufacturing companies will be substantially provided by application software. There are thousands of companies currently supplying such software, and it is to be expected that their product offerings will be extended to support the sort of integration functions that are described in this book. Some of these extensions will be quite straightforward, while others will require implementation of new concepts (as will be discussed in this and the following chapters). However, application packages will not be able to provide complete systems; the users will need to adapt and improve the package functions, and this will require a wide understanding of the integration capabilities of software in general.

A major tool for better decision making, and also for linking different computer systems at the application/user level, is the extended use of database, particularly the flexibility of relational databases and associated query languages.

Distributed database

Data kept on different computers, but controlled as one database, can be managed as described in Chapter 4 (the software structure being illustrated in *Figure 4.5*). A data dictionary would possibly be used to keep track of where the data is held, who may see it and who may update it (the concept of databases discussed in Chapter 4 does not necessitate all the database records being on one computer).

However, distributing data over different 'islands of computerization' as separate databases is a different proposition. *Figure 7.3* illustrates two linked systems: a 'host' system for centralized planning exchanging data with a distributed system in a remote factory (across the road or across the continent). The distributed (or departmental) system could, in fact, be another 'host' computer or a shop floor network.

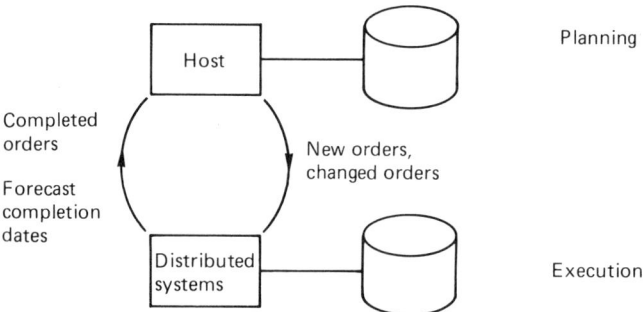

Figure 7.3 Distributed planning and control

It is actually a significant problem to define the subset of the data that should be kept at the remote site, since in practice much of the data will be needed by people at both locations. For example, stock level data is needed centrally for planning decisions (such as whether to accept a short lead time customer order) and at the remote site for issuing stock; the same considerations apply to other data such as engineering change data and customer order details.

If, as a solution to this problem, some of the data is duplicated (i.e. copies at both sites), it is difficult to keep both versions of the data accurate. Even supposing the existence of some fairly major software systems to send transactions from one computer to the other whenever a data field is changed, there will be times when the link between the two computers is down. In such circumstances, the stock level on the host system for instance would not be updated with factory stock movements; sales people on the central computer could be happily promising immediate delivery 'as soon as we are re-connected', unaware that the remote factory has just scrapped a critical batch. When the connection is re-established, a further set of application specific software is therefore needed to check all the transactions that were made during the break to see if any errors have been created and sort out any problems arising. This software would also be used if the different locations were working different hours (e.g. overtime or weekend working).

The application difficulties are much reduced if a logically simple way of segmenting the data happens to exist, for instance the customers of a particular branch. But in typical manufacturing situations, users at the remote location will benefit (even if only occasionally) from being able to enquire around the host data base, and a CIM architecture should allow this to happen. Hence data being distributed over different locations can present significant application problems. However, there are other approaches.

Integration by database

It may be much simpler to leave existing 'islands of computerization' (e.g. shop floor computer, host computer, CAD system) as they are, and link them by setting up additional databases. These additional databases would usually be relational (because of the flexibility of relational databases compared with hierarchical databases, as discussed in Chapter 4.) This approach is illustrated in *Figure 7.4.*

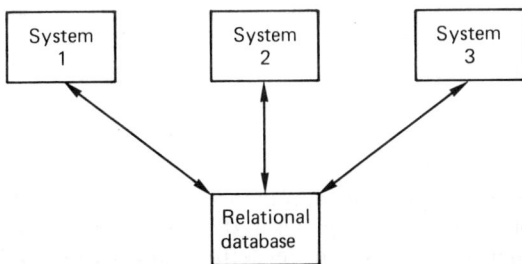

Figure 7.4 Integration of database

Typically, each of the individual 'islands' would retain responsibility for its own data; there would be no doubt who held the 'master' copy of the data. The relational database would contain selected data from the separate systems (the costs of disk storage being sufficiently low that multiple copies in different formats could be accommodated). An example of this approach would be to extract, for control of engineering changes, all the CAD drawings related to a change as well as stock, cost and market data. Always the master would be the original data; effectively the user is taking a chance that the data does not change significantly after it is extracted for decision analysis. (There is nothing to stop the user referring back to the original data, but there is unlikely to be a mechanism for the user to be automatically told if the master data changes.)

Extracting data from one database, storing it on another database, processing it and then updating the original database will be increasingly used in CIM systems. For example, CAD workstations are now capable of extracting a model, processing it and then updating the database with any changes (using say finite element analysis, see Chapter 8). This has the additional advantage of taking large processing loads off the central computer. Another such application might be analysis of capacity availability. Large amounts of data could be extracted from the production database, processed by suitably complex algorithms, and then a judgment taken that a job be subcontracted or an order delayed. The results of this perhaps extensive analysis could then be fed back to the production database by a simple transaction such as a change of date.

SQL (Structured Query Language – introduced in Chapter 4) is important within CIM because of its development as a data interchange language; a program can request data from different databases by generating SQL commands. Hence SQL, which many software packages interface with, becomes the means of transferring data around the company. ANSI has defined a version of SQL.

Message sending
The value of the integrated database will be enhanced if messages can be automatically generated on conditions set by the user. A familiar example is the automatic message to production control when a transaction from the stores produces a negative stock balance; this technique can be extended to allow the user to receive messages generated on any set of circumstances, from delays in the delivery of raw materials to an overload on a key production resource.

The use of integrated computer systems for the sending of conventional messages (i.e. as typed in by the user) will be discussed in Chapter 11. However, an alternative to the integration of databases is to exchange data between them.

Data exchange standards

Data exchange requires the definition and acceptance of common, or neutral, formats that all suppliers can use. In general, these will need to be international standards. If such interfaces can be defined and accepted, then computer systems do not need to be changed, they just need some additional software to unload and load the data. This approach is used particularly in CAD and graphics systems.

CAD data exchange standards
IGES (Initial Graphics Exchange Specification) is probably the best-known graphics interchange system. It was developed with funding under the US Air Force ICAM Programme and the first version of it was published by the National Bureau of Standards in 1980. *Figure 7.5* illustrates the operation of IGES (and other similar exchange standards).

Further versions of IGES have since been produced, with additional features, and most CAD systems provide IGES processors, so that drawings can be exchanged between different CAD systems in the IGES format. Since it is based on 80 character records (as used in punched

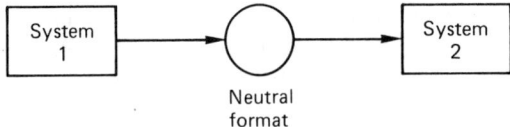

Figure 7.5 Data exchange standards

cards), its performance is less efficient than some users look for. SET (Standard d'Echange et de Transfer) is a major initiative by European Aerospace manufacturers claiming improved performance over IGES.

Several other CAD exchange standards have developed for specific requirements. VDA-FS (Verband des Automobilindustrie Flachen-Schnittstelle) is developed by German car manufacturers to transfer surface data. PDDI (Product Definition Data Interface) extends the transfer of CAD data to include product definition data, including manufacturing process and administrative data. With other interchange formats, including PDES (Product Data Exchange Specification) and XBF (Experimental Boundary File), it is likely to be subsumed into the developing ISO standard called STEP (Standard for Exchange of Product Data).

The focus of the standards in the previous paragraph was on mechanical items. Electronic components have specific and different requirements for which a suitable data format is provided by EDIF (Electronic Design Interchange Format). A useful feature of EDIF is that it represents the data in a format of a computer language (rather like LISP) and can therefore be accessed and modified directly. In other words, if a minor change is required, it can be made directly to the neutral format. This feature is also of value if the software which creates the neutral format is not fully proven.

Hence exchange of data between CAD systems is not straightforward. A major use for these standards is to exchange data between customer and supplier, and some big companies have found a simple solution; they just tell their suppliers which system to buy!

Graphics standards
There are a number of different approaches to graphics standards (in principle, CAD is a form of graphics with some specific requirements; however the standards are separate). GKS (Graphics Kernel System) is an international standard which defines a set of functions for 2-D computer graphics programming (GKS-3D is a 3-D extension to GKS). It is intended for use by applications that produce computer generated pictures. PHIGS (Programmer's Hierarchical Interactive Graphics System) is a 3-D system and, like GKS, is a program interface standard (and is to become an ISO standard). It is designed for highly-interactive graphics environments, including CAD. PHIGS provides and manages an easily modifiable, centralized graphics database. PHIGS and GKS-3D do not compete; they are interfaces at different levels and were designed with different approaches.

PHIGS and GKS define graphic functions in language-independent format, hence additional standards, called 'language bindings', are needed to define how the functions appear in particular programming languages. A benefit of this technique is that different languages (e.g. FORTRAN, Pascal, C) can take advantage of their own data structures and identifiers.

Another form of interface, the Computer Graphics Interface (CGI), defines the abstract functionality of the interface to a family of virtual devices and the behaviour of those devices. As with GKS and PHIGS, other standards will define bindings and encodings of CGI. The Computer Graphics Metafile (CGM) defines a file format for interchange of information. CGM and CGI have a lot of functionality in common.

Like most standards work within CIM, graphics standards are complex and continually developing; it will be useful to relate the standards discussed above to the standards for windowing and for computerized publishing. However, the next section discusses another approach to integration, the development of common operating systems.

Common operating systems

A major candidate for a common multi-user operating system to run on different suppliers' hardware is Unix. The Unix operating system was developed by AT&T Bell Laboratories^{TM} in the 1970s and has been widely licensed; there are now many versions available. Unix is written in C (Chapter 4 introduced operating systems and high-level languages).

Unix includes an OSI-like layered architecture for inter-Unix connections on networks. Proprietary network products will enable Unix systems to be networked with single user DOS-based PCs. (The number of users is important; Unix allows multiple users, DOS on PCs run with one user, while OS/2^{TM}, the second generation PC operating system, is single user and multi-tasking.)

Unix supports a productive programming environment, and the attitude of the software suppliers will be important to the success of Unix (since users choose an application, and not an operating system). Unix systems take more computer resource than DOS systems, and many users may well not want to pay the extra. A further issue is that some applications (e.g. spreadsheets and word processing packages) do not show an interface to the operating system.

OS/2 is likely to provide for many applications a suitable standard operating system. It gives access to multiple concurrent applications for PC users, allowing programs to exceed the 640K memory limitation of the DOS operating system on PCs. Extended editions of OS/2 include a relational database manager and software for communication between PCs and other computers. The database manager includes an SQL data server for application programs and a front end for direct SQL commands by a user. As discussed above, these additional functions will be major tools in the integration of computer systems.

Operating systems are also required for the shop floor network; they will have some different requirements to the systems discussed so far.

Network management systems
Large office networks already exist running tens of thousands of terminals around the world, so a few computers on a shop floor should not

apparently present a problem in operating system terms. However, these large networks are the result of a gradual development of software and hardware over the last fifteen to twenty years, involving thousands of man-years of programming and testing of the software. Much of the software for these networks is for one manufacturer's computers, which is easier than dealing with the wide set of hardware on the shop floor. Furthermore this software is used for customers in all industries, which shares the development cost more widely. The new and specialist environment on the manufacturing shop floor, particularly with the increasing pressure to reduce operator intervention, is going to require extensive software. This is, of course, the major reason for a standardized approach.

However, the ISO/OSI model is concerned with the transmission of files and messages. Network management issues, such as correcting errors or faults in hardware and software, specifying naming conventions so that message destinations are unambiguous, monitoring load on the network and automatically reducing the priority of less-important jobs, are additional functions. Many of the features described in Chapter 4 for operating system, database and communication software need to be included in network control software.

In some respects the shop floor environment should be more manageable because the traffic on the network should be reasonably determined at installation. If more machine tools are added then there is time for the network to be reconfigured. However, unlike office users, robots and machine tools cannot react sensibly to broadcast messages such as 'the system is closing down and will be restarted as soon as possible'.

In summary, there is no clear picture for the development of common operating systems. In an environment where the largest computer in the factory may be a thousand times as powerful as the smallest, this approach is perhaps not likely to give the total solution. Another development is to accept incompatible operating systems, and to define standard interfaces at the user level.

Common user interfaces

Whatever developments take place in the software components of a CIM architecture into the 1990s, the user will be looking for a consistent interface. The user will wish to be guided through the facilities available on his terminal to access the different applications. For the beginner or infrequent user, the guidance will need to be detailed; the experienced user will not want to wait for pages of instructional text to be displayed or to go repeatedly through multiple transactions, hence an 'expert' mode of operation should also be available.

The user should not be required to have a programming background, and should be able to customize parts of the presentation, e.g. use of colour, windowing and volume of voice communication. Some validity checking can be provided on the screen (e.g. field length). Use of program function keys is important to increase productivity, but some element of standardization is required, particularly between different types of terminals. 'Scratch pad' facilities are useful, so that the user can store temporary comments, make quick calculations and move data from one application to another. Message systems will need to have common interfaces. National language flexibility should be included (automatic translation is probably a few years off!).

Security and recovery of data (as discussed in Chapter 4) become significant in defining user interfaces. The need for an expert mode must not allow sign-on and security procedures to be bypassed, and the user should not be able to delete data by accident because the question 'Are you sure you want to delete this data?' could be answered before it was asked. Similar common interfaces will be required for IS systems analysts and programmers.

Software engineering

Tools available for the development of software have gone beyond the structured programming concepts discussed in Chapter 4. An analyst workbench, based on a graphical work station, can provide automated tools for systems analysis and design tasks, using structured graphical languages and methodologies. The Integrated Project Support Environment (IPSE), covering the complete life cycle of software product development, is evolving into a standard as PCTE (Portable Common Tool Environment); there are clearly parallels between the facilities required for the developers of software, and those required for the actual user.

However, this discussion of software integration has not yet considered how to do the analysis required as a basis for better decisions, using the extra data and capabilities that CIM can supply.

Decision support systems

There will be many techniques to assist decision making in CIM. Two stages can perhaps be identified: (1) to computerize judgment decisions which are currently taken by people; (2) to manage the increased complexity of those decisions when much more data is available and a quicker response is required.

The need for a CIM architecture to enable better decisions to be made is all the more urgent because of the continual reduction in the number of people working in a typical factory. Simulation and linear programming, as described in Chapter 6, will be important. Conventional application software, increasingly hardware independent, will become more sophisticated. This is discussed more fully in Chapters 8 to 11. A set of tools which are new within manufacturing industry will be provided by the application of artificial intelligence and expert systems.

Artificial intelligence and expert systems

People are not going to be able to handle all the data available from integrated computers (suppliers' capacity, customers' requests, costs, planned engineering changes etc.). Neither is it going to be possible to specify to the computer systems what should be done in every different circumstance with the precision, say, of the table of tax deductions in a wages system. Computer systems within CIM will be required to make judgmental decisions based on accumulated experience, and this is what expert systems and artificial intelligence (AI) attempt to achieve.

AI is a branch of computer science that studies the use of computers to produce results normally associated with human intelligence. Speech recognition, natural language processing (i.e. understanding ordinary text), computer vision (i.e. scene interpretation) are important fields of AI, and these could all have application in manufacturing industry. AI forms a continuum with control theory, as is discussed in Chapter 10.

Expert systems form another branch of AI, dealing specifically with accumulating human experience (the knowledge base, expressed as 'rules') and making deductions from that knowledge (the inference engine). Accumulated experience is, of course, the basis of human decision making in many fields including industry, and often great effort and expense is devoted to finding an expert or acquiring the relevant experience. To the knowledge base and inference engine must be added input and output modules (knowledge acquisition and intelligent interface) giving the structure illustrated in *Figure 7.6*. The emphasis on 'knowledge' leads to use of Knowledge Based Systems (KBS) or Intelligent Knowledge Based Systems (IKBS) as alternative terms for expert systems.

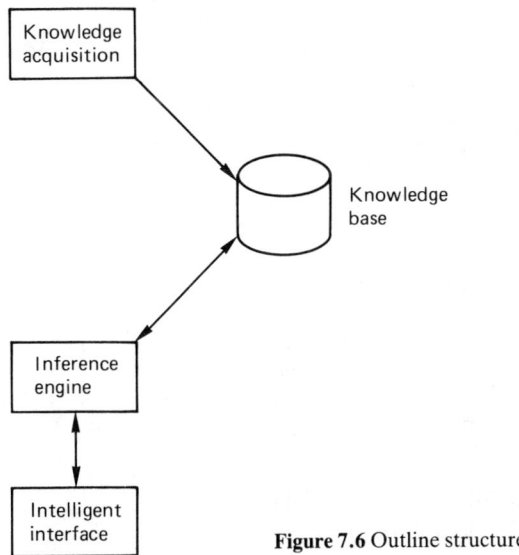

Figure 7.6 Outline structure of an expert system

Expert systems can consist of a 'shell', which is a software package, or programs written in an AI language such as LISP or PROLOG (see below). An expert system shell allows the direct specification of 'rules', e.g. 'If it is summer and if the plant has no leaves then it is 90% certain that the plant is dead'. AI programs will also accept and process suitably structured statements of this type.

A rule applied to plant biology is a typical illustration of an expert system. Many applications of expert systems are oriented to ensuring consistent application of expertise within a known framework, e.g. social security legislation, computer configuration, bank loan procedures. Hence the interpretation of test results is an established application in manufacturing industry (see below); some of the key, never-to-be-repeated, individual decisions that arise in manufacturing industry, e.g. in product design, production scheduling, can also be effectively supported by expert systems.

AI languages

AI programs will be intended to deal with evolving and unpredictable forms of data. This is quite different from a typical scientific calculation in FORTRAN or commercial program in COBOL. AI languages are declarative rather than procedural, a conceptual difference that can be illustrated by the difference between a map (declarative) and a list of instructions (first left, second right, etc.) which is procedural. Another term used in this context is symbolic computing, implying that names rather than numbers are manipulated by the program, though all computer languages except machine code are symbolic to some extent. There are two major AI languages, LISP and PROLOG, with many existing versions of each.

Development of LISP (LISt Processing) started in about 1960. This makes LISP older than most mainstream computing languages except FORTRAN; rather surprising in view of the apparent novelty of computer-based AI as a subject. Whereas FORTRAN has not changed a great deal, so that people who learnt it at college can more or less understand it twenty years later, a major strength of LISP is its extendability. Programming in LISP involves defining new functions. LISP programs and data are both in the same format (i.e. lists), so that one LISP program can be input to another LISP program — try to conceive this in BASIC!. This facility has led to a rich development environment for LISP. Program editors and debugging routines can themselves be written in LISP, which enables the programmer to have a wide control over the language and its use.

PROLOG (PROgramming in LOGic) is another AI language, based on formal facts and their relationships. It was developed in Europe in the 1970s and has been adopted for the much publicized Japanese Fifth Generation Computer Project. The conventional illustration of PROLOG is the family relationship:

parent(fred,david). to define the father to child relationship
brother(peter,fred). to define brothers,

and hence an uncle to nephew relationship can be expressed as a rule based on these two definitions. It would not therefore be necessary to enter every uncle relationship (discussed further in Chapter 10). A question:

?-parent(fred,x).

will return the answer that david is the missing name (PROLOG does not know who is the parent; the programmer or user has to decide, and then enter the data in the correct sequence).

A PROLOG program is able to 'backtrack', i.e. if it has assumed some value for a variable and failed to find a match with the other known facts, it resets the variable to another value and tries again. Also, as a generally accepted requirement for an expert system, it can display the logic by which a conclusion was reached.

AI languages have distinct limitations in their 'intelligence'; if this example were extended:

male(fred,david,peter).
female(fred).

a PROLOG program would not object unless specifically programmed to. In fact, adding knowledge to expert systems about combinations or courses of action that humans would reject (anti-knowledge?) could be a major task.

The example above illustrates that AI programs can be readable, the penalty for which is the need to type in quite lengthy expressions. Needless to say, the programs are unlikely to correct or understand minor spelling errors.

Inevitably PROLOG and LISP are compared. There is little doubt that LISP is the most utilized AI language, and that national (or continental) loyalties can become involved in contrasting their relative merits. From the perspective of CIM, it is useful to spend one or two days learning to use one of them, simply as an educational exercise. The choice between them for a particular application is as likely to depend on availability and convenience of use, as on comparative features.

Expert system shells
For many applications, an expert system shell is likely to be more convenient than writing a program. The mechanics of an expert system shell should not be difficult, about half a day being sufficient to understand the examples provided with a shell and to create a few rules. Syntax of the shell should be in natural language so that programming skills should not be necessary, and the basic tools for editing and screen formatting (both input and output) should have been provided. Reasoning with uncertainty is often provided, for example using Bayesian statistics.

Knowledge-base debugging is an important feature of an expert system shell. If new rules conflict with existing rules or are redundant, then the user will want to know. Under certain circumstances, missing rules could be identified by syntax checking. The shell should be able to display its rules and explain any conclusions it has reached, called forward and backward chaining.

An expert system shell should allow interfaces to be written to access other systems if required. Often there are two sets of software, for developing the expert system and for enquiring on it. The software for the latter is simpler and cheaper.

The optimum process for obtaining knowledge for an expert system is not clear; whether it will actually be done on an abstract basis by a 'knowledge engineer', as is frequently advocated, will doubtless be established in due course. In order to apply an expert system, the logical process of the application has to be understood and documented, probably on a flow chart or block diagram basis. Frame-based representation is a structured format for the representation of knowledge so that it can be understood by the expert and by the computer system. It is often useful to simulate 'consultations' with the expert system to check on the format of the input and output.

A classic application of an expert system is the evaluation of test results; the matrix of known results might be as shown in *Figure 7.7*, though the problem would need to be more complex to justify use of an expert system.

	Test 1	Test 2	Test 3	Test 4	Test 5	Test 6
Fault A	P	P	P	F	P	P
B	P	F	P	P	F	P
C	F	P	P	F	P	P
D	P	P	F	P	P	F
E	P	P	F	P	P	P

Figure 7.7 Matrix of test results

This example illustrates some of the limitations of expert systems. Firstly, AI languages are not essential for the development of a computer program into which test results can be fed (passed Test 1, failed Test 2, etc.) and the fault diagnosed, since such a program could be written in many other languages. Secondly, the use of an expert system introduces questions about the meaning of the data. Would it be reasonable for instance to deduce from *Figure 7.7* that a product which fails Test 1 therefore has Fault C? Expert systems would be likely to make this judgment, which would be reasonable working only on the data in the table, whilst the engineer might insist that all the other tests be completed in case there is a new type of error. Are Faults D and E in fact the same, i.e. if Test 3 fails, does the result of Test 6 have any significance? This is an engineering judgment that has to be fed directly into the rules; the expert system cannot be allowed to make a judgment.

Hence the vision of the 'knowledge engineer' (perhaps a psychologist) interviewing the crotchety but experienced test engineer, who is incapable in some scenarios of understanding his own thought processes, in order to record how problems are tackled and then replicate the process with an expert system, begins to look improbable. This particular example requires a program to be built up by rigorous engineering decision; the choice of programming language may well be secondary.

This may leave expert systems where some people see them, as a programming productivity aid and a flexible tool for developing prototype software. However, they are at an early stage of development, and whether it is called CIM or not, manufacturing companies still have the real problem of massive amounts of data, not necessarily complete or totally accurate, many decisions to be taken and the need for some type of computer based judgment; AI and expert systems almost by definition offer the best hope. Further applications of expert systems are continually being developed; some common examples are discussed in Chapters 8, 9 and 10.

Conclusions

Conventionally, computer systems in manufacturing industry could be reasonably divided into mainframe and shop floor systems. The differences between the two systems (as discussed in Chapters 4 and 5) were seen in function of the application, the computer hardware and particularly the software. The mainframe computer typically ran high-level language programs doing business planning functions such as MRP, a criterion of a mainframe in this context being the capability of running a major manufacturing database (including multiple logical relationships between items, bills of material, suppliers and customers etc.). The shop floor computer was often programmed in an assembler language; it performed data collection (i.e. of production quantities) plus process control. Direct links between these two types of systems, if they existed, were usually limited to file transfer (i.e. periodic transfer of files of data, as opposed to continuous interactive transactions).

Evolving from these systems is the CIM hardware architecture that was shown in *Figure 7.1*. The interpretation of this in application software terms (i.e. the user's) is a planning hierarchy. As described in Chapter 5, the Master Production Schedule Planning functions perform long-term planning, after which MRP produces a detailed material schedule and from this is determined the work-to lists of the shop floor. The planning could be all done on the mainframe computer (or computers); the shop floor network would then be used only to issue instructions. Whether the detailed decisions are taken by the mainframe computer and then explicit commands sent to each individual machine, or whether the work is parcelled out to shop floor network systems to make final detailed decisions is not a matter of principle, but of best business results. The analogy with conventional (i.e. people based) factory organization is obvious — how much autonomy is allowed to a particular factory or foreman is largely determined by the nature of the business, the style of management and the current availability of planning tools.

Hence the question arises of how many levels would be appropriate for the hardware hierarchy of *Figure 7.1*, and the possibility of a hierarchy with four or even five levels. The apparent neatness of the multiple levels overlooked the fact that much of the data would be common to the different functions. As described in this chapter, maintaining multiple overlapping databases accurately is a classically difficult exercise. Neither does this multi-level model take sufficient account of the reductions in timescales and numbers of people which is taking place in manufacturing industry.

No easy solution has emerged. Many international standards are required. Networks are needed for the shop floor, and MAP, based on the OSI reference model, is the outstanding contender. Common operating systems will provide some basic hardware independence, and other software interfaces are becoming available. CIM application software, including integrated financial and costing systems, will need to be highly flexible. AI and expert systems are likely to be required for dealing with the excess of data now available.

Relational databases are important; particularly they allow a user considerable flexibility to develop solutions to manufacturing problems using the data available on integrated computers. The architecture should permit a user to do the job with the most appropriate data in the most convenient way.

The (necessarily) slow process of international standards will require a lot of intermediate judgments to be made about CIM and manufacturing strategy before widely accepted standard solutions can be available. These decisions will require either that senior people have a good understanding of the issues involved, or that some strategic and expensive decisions are delegated by default to technical specialists.

The approach to CIM Architecture under development by the ESPRIT CIM project is discussed in Chapter 13. However, any CIM architecture has value only if it benefits manufacturing businesses; the actual applications of CIM need therefore to be explored further in the following chapters.

Further reading

CARRINGER, R. A. (1985) *Integrating Manufacturing Systems Using the Product Definition Data Interface*, Proc. of Autofact '85, Detroit

ENDERLE, G., KANSY, K. and PFAFF, G. (1987) *Computer Graphics Programming, GKS — The Graphics Standard*, Berlin: Springer Verlag

FORSYTH, R. (1984) *Expert Systems — Principle and Case Studies*, London: Chapman and Hall

GEVARTER, W. B. (1985) *Intelligent Machines — An Introductory Perspective of Artificial Intelligence and Robots*, Englewood Cliffs, NJ: Prentice-Hall

GERSTENFELD, A., GANZ, C. and SATA, T. (1987) *Manufacturing Research Perspectives: USA–Japan,* Elsevier: Amsterdam

MAP/TOP Interface, Society of Manufacturing Engineers

McFARLAN, F. W. and McKENNEY, J. L. (1983) *Corporate Information Systems Management — The Issues Facing Senior Executives*, Homewood, Ill.: Richard D. Irwin

NAYLOR, C. (1983) *Build Your Own Expert System*, Chichester: Wiley

Chapter 8

Product and process design for CIM

Introduction

The design process varies from industry to industry; it also takes different forms in response to the product or system to be developed and the resources that are available. For example, a large piece of capital equipment may take a long time to design and develop and consume large resources in terms of manpower and computer support. In other instances, say for an innovation in an electronic device for the technical market, the preliminary design may be scant and a first rudimentary design may move rapidly into a breadboard prototype for testing and development. As a consequence, little theoretical analysis may be used and the majority of the product design and development would take place at the prototype stage.

However, in spite of apparent differences, there remains a consistent sequential progression at the heart of most design processes; only the magnitude of effort, emphasis, and techniques differ. This sequence can be expressed simply in the schematic diagram shown in *Figure 8.1*.

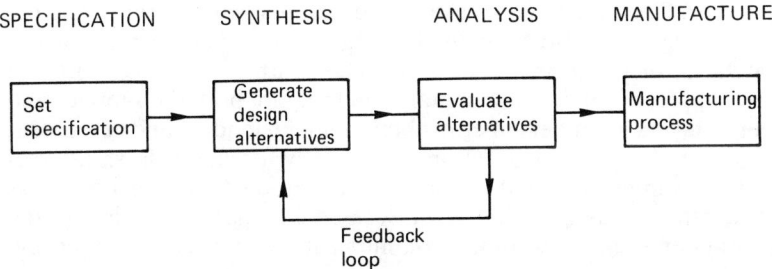

Figure 8.1 Sequential progression in the design process

The design process

The essential first step is the setting of the design specification. Designers need to work within product and process limitations and often seek such constraints in a process referred to as 'closing the problem'. A part of the exercise of closing the problem is setting the specification. This stage of design must also consider business strategy as discussed in Chapter 2. For the purposes of this chapter it will be considered that the business aspects of a venture have been satisfied, and the company has decided to take the design process on through to the establishment of a product specification.

Several tools are available to the designer at this stage. The most important design criteria are those set by the potential market, the manufacturing facilities available and any legal or standardization constraints. The standardization condition may be set externally by safety bodies, standards institutes or by industry-wide agreement. In many other instances there are internal constraints based on the existence of similar products that the current design is intended to supplement or supplant.

The information about existing products, processes and market potential can be made more accessible and usable by integrated computer-based systems. An essential advantage of such systems is their ability to create new designs by modification of existing designs. For example, an existing product may already be described within a CAD database and in this case the designer simply calls up the existing design and modifies it to meet his current requirements. Much of engineering design follows this line and the practice of 'subbing' a drawing to produce a new design is institutionalized in the procedures of many Drawing Offices and their archives. Possibly the most significant contribution made by the computer to the design process is its ability to communicate and modify designs with speed and efficiency. However, CAD and its integration with databases and other information systems goes beyond simply revising a drawing.

Where the designer is highly constrained by formal safety standards, it is common to parameterize designs so that the designer calls up computer based standard design elements that meet these standards, and then works within this framework to create a new design.

The first attempt at setting a design specification is often in response to a customer request for a quotation of price and delivery for a piece of equipment. If the equipment to be supplied is in any way customized, the design process must start with establishing agreement between the supplier and customer. The agreement is usually based upon some form of written contract, which may also include drawings, costs, conditional clauses, delivery dates, or progress breakdowns. The contracts may be short and relatively simple; often they are complex and require both the supplier and customer to take a great deal of effort before understanding and agreement can be reached. Such negotiations frequently occur with short real deadlines. It is important that the setting of a product specification be as quick and efficient as possible. As discussed in Chapter 11, integrated computer systems can assist in the communication between customer and supplier.

To establish an efficient and reliable quotation (or tendering) scheme the products and services a company wishes to offer must be carefully defined. A company offering standardized products will have no special design requirement to respond to a customer request. For non-standardized products and equipment, the later stages of design and the setting of specifications interact. Information, skill and intuition are often required to reach an acceptable specification for both sides.

Another major advantage of the use of a computer system is the high degree of detail and discipline required. There should be no room for vagueness at the specification stage of a design if the design is to progress and eventually reach efficient manufacture. The use of computer integration helps to impose that desirable discipline.

Up to the present, technology has produced few generally applicable tools to help the designer to set a specification, although the use of product and processes databases in the form of CAD and Computer Aided Process Planning (CAPP) packages has offered some assistance.

In addition, there has been little practical progress in the use of the computer to aid the engineer in the more innovative stages of design. Research by the AI community, mainly in the USA, has attempted to use the computer to mimic the inspirational designer, but on occasions when both practical and academic goals have been combined, it has been necessary to severely constrain the product design possibilities. However, the implications of the combination of AI and design are far reaching, and there is much potential for progress even though the task is difficult.

If the inspirational designer sees the specification stage as establishing the reference framework for the subsequent design, it is the synthesis stage at which imagination, experience, flair and skill come to the forefront. During this stage the designer or design team must establish the details that define an acceptable product in what can be seen as the heart of the design process. Often the product will consist of materials shaped to form components that are in turn fabricated into subassemblies and assembled in final products. As a consequence the designer will be concerned with material selection and the shape that material must have to achieve its function. In the past this creative action was performed during the migration of ideas between a note pad and a drawing board. Such migrations were often inefficient and the production of orthogonal or isometric drawings was never more than a clumsy tool for expressing the shape of three-dimensional objects. The use of 2 or 2.5 D drawing packages may be adequate for many applications but, increasingly, full 3D modelling of solid shapes or surface contours are being used. Even with quickly created conceptual models, valuable information can be established. The use of CAD offers improved design visualization tools in addition to freeing the designer from much of the drudgery of repetitive drawing, design detailing and dimensioning. The experienced user of a CAD facility is four or five times faster than his drawing board based counterpart, typically reducing the total time for design and manufacture by something like 30%. Often companies have found that the introduction of CAD has allowed their designers to produce more complex designs than can be achieved on a drawing board. As shown in *Figure 8.1*, feedback loops occur in the design process if the customer requirements have not been properly specified or deficiencies are found in the initial design proposals. As the synthesis of the design begins, a CAD system can be used to build computer models. The power and flexibility offered to the designer at this stage allows specification and design concepts to be a highly interactive and controlled process.

The initial part of the synthesis process, although aided by the use of the computer from the point of visualization, is relatively little changed by CAD but speed and convenience probably leads to more design concepts being investigated and in more depth.

As the synthesis stage progresses, the designer is able to continue refining the most promising conceptual designs in a process of synthesis and analysis. This is the most demanding of the stages in creating a design and needs to be given more detailed consideration than other stages. During this stage of design, opposing demands may be placed upon the designer. It is necessary to meet the agreed specification in function, price, delivery, durability, esthetics and other factors imposed by the customer. In addition, the designer must meet the business requirements of the designer's company. This will primarily be related to the efficiency of the design process and the subsequent manufacture.

Often, constraints are set by the manufacturing facilities that are available. The eventual efficacy of the design can depend upon how accurate a conceptual model of the available manufacturing facilities is available. Historically, in the manufacturing industries, designers were recruited from the shop floor and they brought into the design office an invaluable knowledge, not only of general methods of manufacture but a very specific knowledge of the machines that were available on the shop floor of their own factory.

Due to technological and social changes, this system of knowledge transfer by movement of staff from manufacture to design is being eroded and in many cases it fails to meet the demands of increased complexity and efficiency. For example, motor vehicles are no longer designed and manufactured on a single site but may be designed on one side of the world and manufactured on several sites on other continents. Under these conditions a designer needs to know what is available within manufacturing plant; this information can be made available by an integrated system. Similarly, in high technology areas, the designer is required to understand complex and powerful design tools that require specialization in their use. It is not possible for a designer of high technology products to have started work on the shop floor and 'worked his way up'.

As the designer refines the design, tools such as finite-element packages may be used. Eventually the design will be completed such that it gives a match to the specification and completely defines the geometry, material and process requirements for creating the product. When the design has reached this target the detail designer or draftsman will produce the documents required for manufacture. The design function may extend to the preparation of process plans, part programs for NC machine tools or robot programs.

Having considered the design process in a general sense, it is now appropriate to move to more detail on the systems used.

Use of CAD systems

CAD is little over twenty years old and the credit for its origins is usually given to the SKETCHPAD system developed at the Massachusetts Institutes of Technology. The fundamental function of CAD is to represent the form of an object to allow its creation, analysis, synthesis and communication. Considerable advances have been made over the last twenty years, as exemplified by the shift from 2–2.5 D drafting packages to full 3D modellers. However CAD system designers are continually seeking to develop better representations using the latest available workstation hardware and technology.

It will be useful to see how geometric elements such as points, lines and circles are combined within a computer-based numerical environment. An obvious and familiar tool is the use of coordinate geometry, in which geometrical features are expressed with reference to a grid system of rectangular (or cartesian) coordinates. For example, a line can be defined in length and position simply by stating the cartesian coordinates of its end points. Such a definition will often define a feature uniquely and enable it to be transmitted into any identical coordinate framework system. However, a line defined by its end points and a knowledge of the coordinate system is not completely defined. In order to fully define the line, extra coded information is required to describe the line thickness, the line type and colour. If a sequential code is used, then the line may be defined by a string of numbers, where portions of the string represent different line characteristics. For example a line could be represented by 00358491 where the string has the meaning shown below:

Geometric element	0	(this means a straight line)
Coordinates of ends	0,3–5,8	(first point is $x=0$ $y=3$ and second point is $x=5$ $y=8$)
Line type	4	(this means a chain-dot line)
Colour	9	(this means green)
Line weight	1	(this means 0.1 mm thick)

In practice geometric features in CAD systems are somewhat more complex and the string of digits is longer than the example.

As a consequence of the ability to specify a geometric element as a string of numbers it is possible to specify a whole drawing as a file containing such numbers.

The direct use of coordinate geometry at such a fundamental level would be slow and unwieldy. Fortunately computer hardware and software has been developed that allows a designer to use this technique in an easy manner. However, the ease with which a designer is able to create and manipulate these geometric entities is not only dependent upon hardware but is considerably affected by the special software tools or programs that have been developed by CAD system designers over the last twenty years. This software takes the form of CAD packages such as CADAM and CATIA and many others. In the main, packages that are used in industry are large and powerful, initially running on mainframe computers but more recently moving to the new generation of powerful workstations. Systems of this size represent a major investment for both the package producer and user alike. Other systems, used mainly for drafting work are based upon PC hardware, for example AUTOCAD, and these lower cost packages are being adopted widely by smaller and low technology companies.

The problem faced by the CAD system designer is gaining a meaningful representation of the attributes and presenting it to the designer for his application. Most users are unaware of the way in which the package is organized and its internal functions but the external appearance of the software is of considerable interest and importance to the user especially the manner in which the CAD system collects its data and how the data is displayed. The illustration in *Figure 8.2* shows a typical CAD screen for inputting geometrical data. The technology of the man–machine interface in CAD is continuously improving but the basic methods of communication and interaction are now well established for both input and output of geometric data and have been described in Chapter 4.

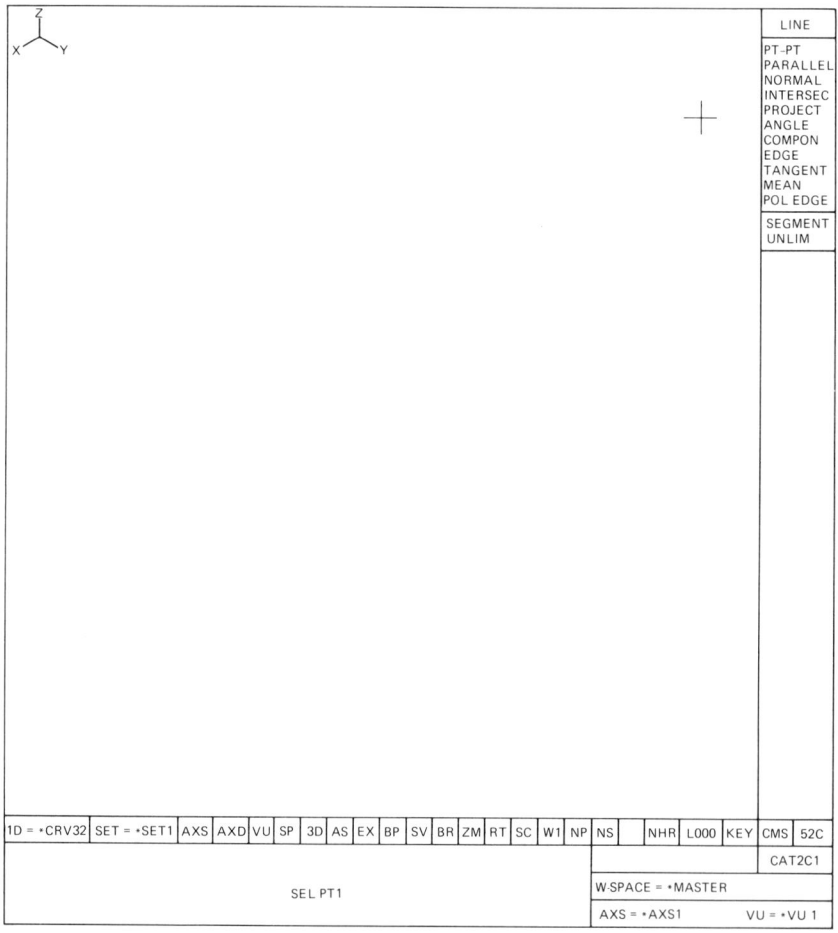

Figure 8.2 CAD screen with menus (top right) for inputting geometrical data

A concept that increases the convenience of CAD representation is that of parameterized design. In this technique the object being designed or modelled is considered to have a general form. Each individual object can be distinguished from others, and fully specified, by the identification of that general form or structure and its individual parameter values. As a simple example, a rectangular box could be specified in general terms as 'rectangular box' and its size defined by stating its height, width, and

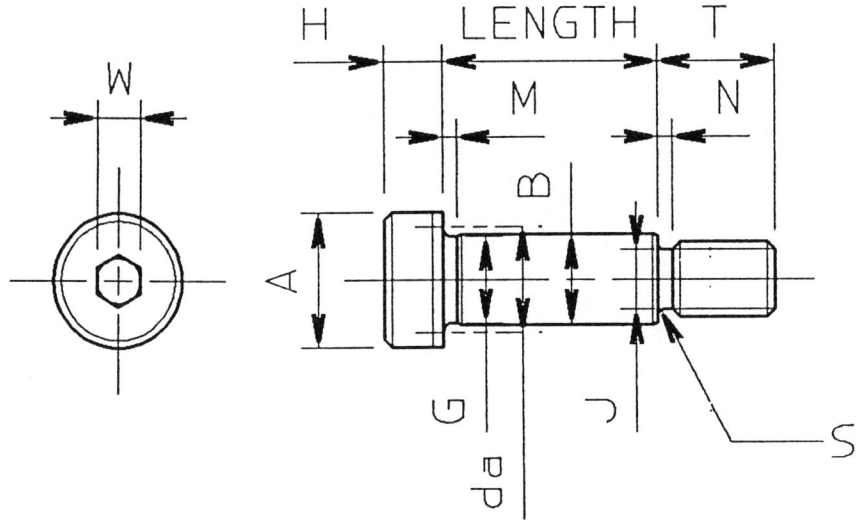

Figure 8.3 Parameterized design of a shoulder bolt

depth. Applying this basic concept to more complex objects requires an extensive range of parameters such as shape, size, colour, material and so on. An example of a parameterized design is shown in *Figure 8.3*. The example shown is a shoulder bolt, a commonly used fastener. In this case the CAD package stores the design of the part in parameterized form so that the designer is only required to specify the values of the parameters, for example length or diameter, for the software to generate and display the detailed part design. Parameterized design databases are useful in many areas of design, but are particularly powerful and effective in areas where design features are standardized and used reparatively such as electrical circuits and pipework and pressure vessels for process plant.

Although CAD systems usually recognize and represent a wide range of attributes, it is the geometrical representation of shape that predominates.

3D representation

There are three widely applied forms of representation of the shape attribute of an object, namely:

Wire frame models
Surface models
Solid models

The differences between these three representation techniques are illustrated in *Figure 8.4*.

The differences can be considered in geometrical terms. As its name suggests, a wire frame model is a representation of shape using lines to represent edges of the shape. This is quite adequate to represent the shape, position and orientation of an object but it is inadequate in many instances

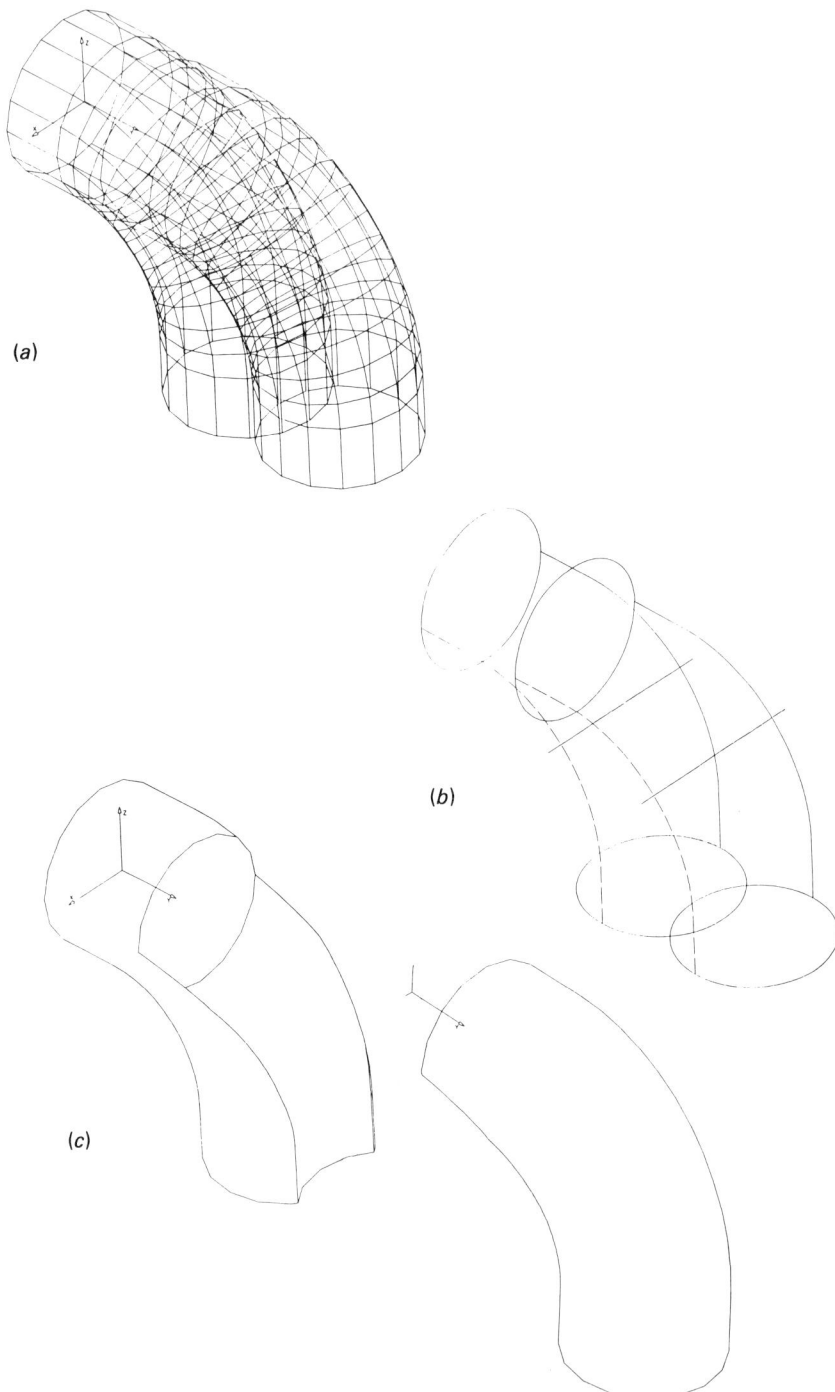

Figure 8.4 Representation techniques in CAD: (*a*) wire frame model (intersecting pipe bends), (*b*) surface model, (*c*) solid model (intersecting pipe bends moved apart)

to represent the interaction of different objects or complex surfaces. In these cases it may be more appropriate to use a model that defines surfaces of the object. In surface models the object shape is represented by face surfaces and bounding edge lines and corner points, which is particularly useful for the creation of surfaces through an NC machining process.

Lastly solid models are defined in terms of corner points, edge lines, face surfaces and internal volumes, a collection of parameters that lends itself to the determination of weights, inertias and other dynamic properties. The solid model is useful for dealing with problems of interference or interaction of two objects. Solid models are usually built up by combining a number of solid primitives in a technique known as constructive solid geometry. The ability to add and subtract one solid with respect to another is a very powerful design facility.

The interpenetration of two cylindrical elbows is a useful example to show the differences between the above representations. In the first case, that of the wire frame model, the representation is unable to infer anything about the shape of the edge created between the two cylinders nor can it define the shape of the interface surface; the two wire framed models are simply interlinked, see *Figure 8.4(a)*. The surface model is much more useful than a wire frame model in that its description of surfaces allows it to compute and consequently represent edges created when surfaces meet as illustrated in *Figure 8.4(b)*. As a consequence, a surface modeller could be used to determine the shape of the surfaces excluding the shared surface and to use that data, for example, to obtain the surface development required to make the combined cylinder as sheet metal parts. However, although surface modellers are frequently used, they carry no information about the contained volumes and therefore are not able to make inferences based upon the knowledge that a point is within a body or outside it. The drawing in *Figure 8.4(c)* illustrates the effect of subtracting one elbow from another where both elbows have a partial shared volume.

Dimensioning and tolerancing

Communication between designer and manufacturer usually takes the form of a 2–2.5 D drawing that may be produced directly by the designer using a drafting package or may be derived by the designer from 3 D design models. An essential element of the information carried by the drawing is associated with dimensioning.

In the design process using CAD, the designer may develop ideas of shape by various techniques. In essence the CAD system keeps a record of the geometric elements and their parameters as has already been described. As a consequence, it already holds the information required for the shape of the product to be generated in manufacture. However, the CAD database will contain more information than is necessary for specifying the part to be manufactured and so one of the last tasks of the designer is to indicate a minimum subset of dimensions that the CAD package needs to display in order to allow manufacture.

The term 'automatic dimensioning' is sometimes used to describe the function of adding dimensions to a drawing within CAD and can be misunderstood. As it is inappropriate to display all dimensions available from the geometric database, the designer must indicate not only which dimensions are to be displayed for manufacture but also how they are to be displayed. The 'automatic' aspect of the system occurs when the designer has made these choices, at which point the CAD system can automatically insert the dimension lines and the appropriate distance value adjacent or along the dimension line. Hence, although human error can lead to a CAD drawing being under dimensioned or containing redundant dimensions, incorrectly specifying the magnitude of a dimension should be virtually eliminated.

In some CAD packages the style of dimensioning is set, in others the designer has the freedom of specifying a number of global parameters. For example, it should be possible for the designer or design system office manager to specify such parameters as witness line thickness, text size, dimension value position relative to the dimension line, arrow heads etc. Once the general style of dimensioning has been set, usually at office level, it still remains for individual designers to make decisions relating to a particular drawing.

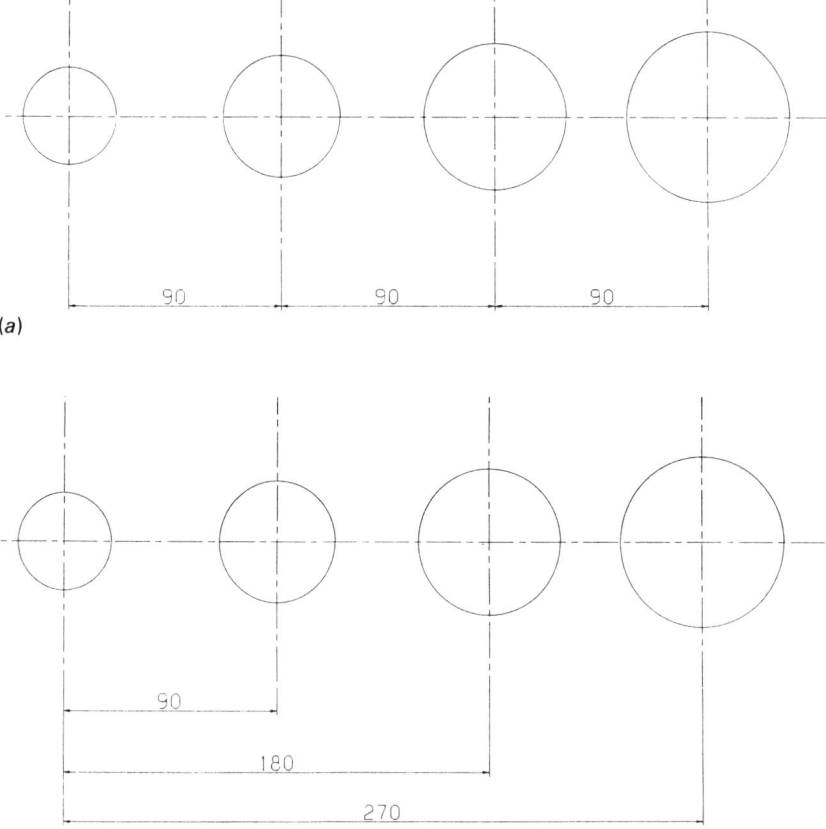

(a)

(b)

Figure 8.5 Dimension string and dimension stack: (a) dimension string, (b) dimension stack

There are two types of dimension in a drawing, linear and angular. In linear dimensioning the magnitude relates to the distance between two points or lines joined by a straight line. It is usual practice to string or stack dimensions as illustrated in *Figure 8.5*. Note that the stacking of dimensions avoids the accumulation of errors by referring all dimensions back to a single datum.

The procedure for adding dimensions varies but essentially once the dimensioning and distance functions have been selected, the designer has only to indicate the end points or lines of the dimension. It is usual for the designer to specify if a dimension is measured horizontally, vertically, or at an angle. Finally the designer will be asked to indicate where he wishes the dimension to appear. Tolerances can usually be added to a dimension by specifying upper and lower limits on the computer held nominal distance. The illustration in *Figure 8.6* shows a typical screen display for dimensioning.

The specifying of angular dimensions follows very similar lines but in the

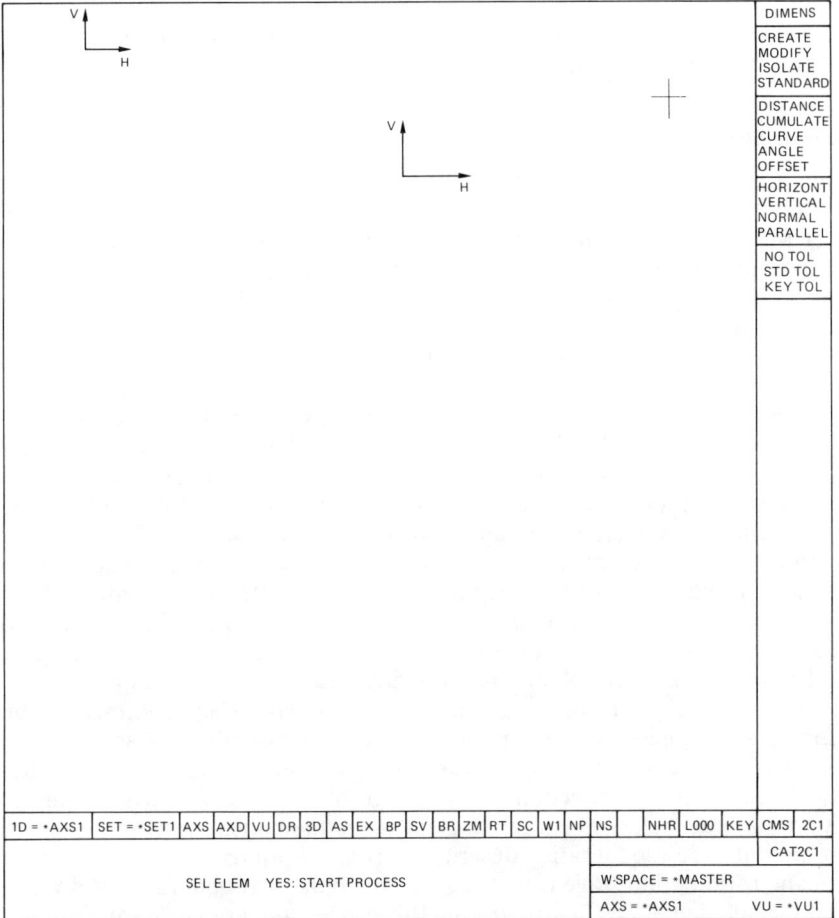

Figure 8.6 CAD screen for dimensioning

case of angles the designer must specify the limiting lines of the angle and indicate where the arc of the dimension line and text for the magnitude should appear.

Circles and arcs of circles are common features in most mechanical designs and as a consequence their dimensioning usually has a special procedure. Usually the designer only has to indicate the line of the arc and where the dimension should appear. In most systems it is possible to toggle between specifying radius and diameter.

In addition to the standard dimensioning facilities, it is also possible to add special nonstandard dimensions using text, lines and arrows manipulated and specified via a mouse or keyboard etc.

Analysis packages within CAD

Significant advantages of CAD arise from the possibility of using a CAD database as the input to powerful computer-based analysis tools such as finite element packages. Two important but different examples will be considered; the finite element methods for engineering analysis and electronic design and network layout.

Finite element methods

Finite element methods are typically used to gain approximate descriptions of how an unknown value, for example stress, varies throughout a finite region or domain within a component. In the finite element method, the component is divided into elements over which the unknown is considered to vary in a known continuous manner. Typically the elements are equivalent to simple elements such as beams, shells or rods, for which classical theory can be applied. At the interface between the elements, the transfer across the boundary will not be smooth but the technique is to seek the best possible continuous function that will satisfy the end conditions. Clearly, under complex loading of an irregular shaped part, the stress field will vary considerably and as a consequence a good approximate solution will rely upon repeated calculations over a large number of small elements. The density of the elements can be varied and the designer will usually use a fine mesh of small elements in regions where the unknown is expected to change rapidly. Since the early application of finite element methods in solid mechanics and structural analysis, the approach has been extended and extensively used for flow analysis, heat transfer analysis, and many other areas where problems can be expressed in the appropriate form.

The finite element method consists of four basic stages, known as the formulation phase, the evaluation phase, the assembly phase, and the solution phase. These stages can be discussed more fully using the determination of deflections in a structure as an illustrative example. Various other texts give a more detailed mathematical description of the topic but here a qualitative description is appropriate.

The formulation stage is the stage in which the part under load is divided into small elements. Clearly, during this process the graphical capability of a CAD package is not only useful for supplying the description of the part design, but it also provides the means of generating, storing and displaying the grid system used for subdivision. The diagram in *Figure 8.7* shows a finite element grid applied to an automobile wheel.

As a simple example of the finite element technique, the deflection of a simple beam under an irregular load could be considered. Subdivision into elements could be arbitrary; simply dividing the whole into a convenient number of short lengths. In other cases, the subdivision will be on the basis of nodes placed at points of some physically significant condition such as a point of application of a load, or a change in beam section.

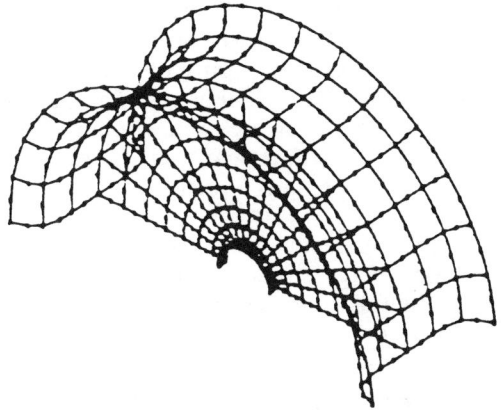

Figure 8.7 Finite element grid applied to an automobile wheel

Building up the finite element model for a complex component is usually an iterative process in which an initially crude model with few elements is refined using more small elements at critical points within the model. During this process there must be a compromise between the accuracy and confidence in the solution and the computing time consumed, a compromise that changes year by year as computing becomes cheaper. Within solid mechanics, the models are often a framework of simple beams, each beam being assigned the usual material properties for example Young's modulus, bulk modulus and structural characteristics such as section modulus. More complex components may be modelled using isoparametric elements. These elements not only have nodes at corners, they also have additional nodes at the midsides, a technique that reduces the amount of computation required without loss of accuracy. Isoparametric elements can be used to model axisymmetric, plane stress, plain strain loading of flat plate, thin shells, thick shells and solid blocks.

Symmetrical objects need not be fully modelled and interfaces at axes of symmetry can be dealt with by use of reflection techniques producing identical fields across interfaces.

Some time ago it was necessary to input finite element models by laborious manual input but modern systems are fully integrated into CAD packages with automatic mesh generation routines.

In addition to the generation of this predominantly geometric data, the formulation stage includes the specification of the manner in which the unknown varies. Using the beam example, the unknown would typically be its transverse deflection under load and the manner in which the deflection varied from nodal point to nodal point could be linear or based on a quadratic function. In this stage the system user is able to manipulate the assumptions to produce solutions under different conditions. For example, it could be assumed that the deflection within an element was linear and the inaccuracy introduced by such a coarse assumptions could be compensated by the use of a large number of relatively small elements.

Within the evaluation phase, the program determines the contribution of each of the elements towards the final total condition. This is more clearly understood by use of the beam deflection example. In this case, the overall condition to be minimized is the strain energy stored in the beam for a particular beam deflection. Thus, evaluation consists of expressing the strain energy in each element of the beam in terms of the stresses generated in the beam and the movement of the nodal points. This part of the analysis leads to the formulation of an element stiffness matrix that contains all the information to specify the stiffness of each element. In addition, a second term is produced that appears as a vector describing the force acting on the element.

During the next phase each of the elemental strain energies is added together to produce the total strain energy stored by the system. This process produces a large system of equations that describe a global stiffness matrix and a global force vector that correspond to the whole beam in a similar manner to the elemental terms described in the previous paragraph.

In the final phase, the solution phase, the unknown values in the equations that were derived earlier are solved. It is the solution of the large number of simultaneous equations in this phase that requires the calculating speed of the computer. In the case of the example, the unknowns would be the deflection at each of the nodal points. The basis of the analysis minimizes the strain energy in the beam. To be more specific, the process of minimizing the strain energy is performed by the partial differentiation of the energy term with respect to nodal deflections and equating to zero. Part of this solution phase requires the input of known or assumed boundary conditions that are set by the physical constraints on the system. For example, the ends of a beam may be constrained from deflecting or rotating.

One of the most difficult tasks in the use of finite element analysis is the interpretation of results. Results from this form of analysis are more easily visualized in graphical form and in recent years there has been increasing use of colour graphics to reveal to the designer the zones of high stress. This display generation process is known as graphics post-processing and allows large amounts of data to be fed back to the designer in an easily digestible form.

In industries where the analysis of stress is an important aspect of design the integration of CAD and finite element analysis is now a well established tool.

Electronic design and circuit layout

The design of electronic circuits consists of two interlinked phases: the logic design that defines the logical circuit required to change from input to output within the device, and the physical design that is the physical layout and interconnection on a board or within an integrated circuit. In recent years, these two phases have led to the development of two types of computer aids. The first to be developed were tools to help the electronic layout designer to deal with the physical positioning of electronic components and their electrical connection. This type of package is often known as CAED, computer aided electronic design. At its most basic level, it represents a facility for drafting schematic diagrams, logic networks, the geometrical layout of components and the copper track design for component interconnection. More advanced packages allow some system checking, prepare connection lists, photoplot for PCB mask manufacture, and prepare NC drill tapes and automatic part insertion programs for PCB stuffing machines. Although somewhat specialized, the manufacture of PCBs represents a good example of CIM.

A PCB is a sheet of insulating material that carries an etched pattern of copper tracks. The copper tracks connect the electronic devices, which may be mounted directly onto the surface of the board using a conducting adhesive or more traditionally are mounted via leads that are soldered into holes predrilled into the board. Boards may be single sided, double sided, or multilayered to achieve the desired interconnection between circuit components.

The layout of a PCB using a CAED package can be carried out either automatically or interactively with the designer. Most modern systems allow both options, and the user selects the most appropriate approach depending upon the complexity of the circuit. There are three basic stages

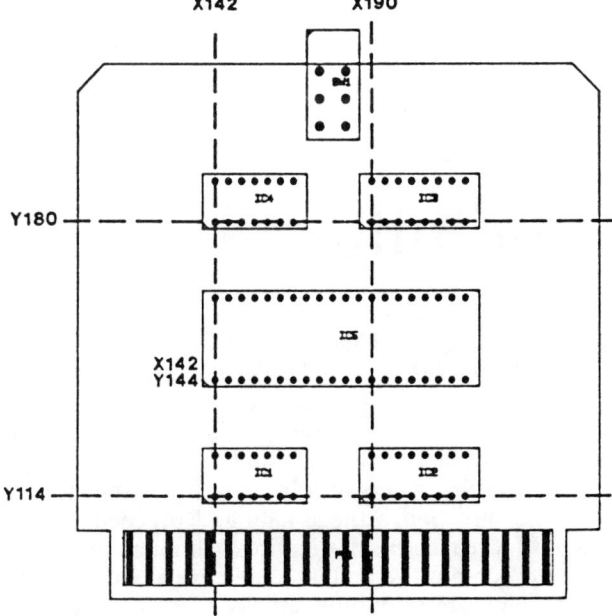

Figure 8.8 Physical layout for the PCB

in the layout process: the placement of the parts, the initial design of the interconnections and the final design of track routings.

Component placement on the board is critical because it exercises a considerable influence upon the subsequent track design. The aim in this process is to devise a layout that is compact, with the minimum track length, but still allows all of the desired interconnections to be made. In the interactive approach, the designer decides upon the placing of the components using experience and basing decisions upon physical constraints of size and thermal considerations. The illustration in *Figure 8.8* shows a physical layout for a PCB circuit.

In the second stage the designer defines the interconnection between the individual components. A technique known as 'rubber banding' is useful at this stage. This technique allows the designer to display the connections made between each component on the VDU. During the development of the layout, the designer is able to move the components about the screen and the connections are maintained by a stretching of the interconnecting lines. The diagram in *Figure 8.9* shows a typical connection diagram.

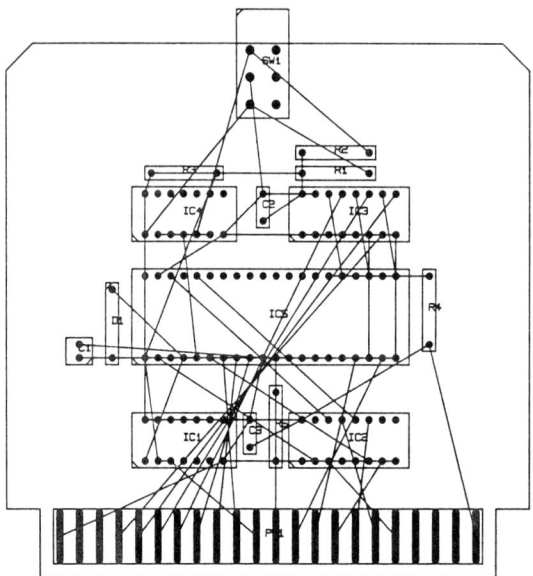

Figure 8.9 PCB connection diagram

The final stage, that of designing the final route, is the most amenable to the use of computer aids. Routes that give the desired interconnection without violating any of the design constraints can be devised using routing algorithms. The quickest and simplest of these is a 'line search'. In this algorithm the route starts at the first component and moves in the x or y direction towards its target until it meets an obstacle. Upon meeting an obstacle the circuit changes direction orthogonally and proceeds to the next obstacle and so on until it reaches its destination. Although quick and simple this algorithm may not always find a route, even when one is possible and the routes found may not be efficient or compact. An algorithm that is always able to find a route but may require considerable computing time is Lee's algorithm. In this algorithm, the whole of the layout area is divided into small squares. Each square in contact with the source is identified and marked with a 1. All unmarked squares in contact with squares marked 1 are marked 2 and so on. This process is continued until the destination is reached. All of the potential routes are thus identified as a series of sequential numbers starting at the source and ending at the destination.

Even though commercially available CAED packages usually use several different routing algorithms to arrive at a satisfactory layout design, it is not unusual for a designer to find that the part placement chosen prohibits the design of a satisfactory route layout. Hence, there is a frequent requirement for the designer to interact with the computer to tidy up the design and bring about an acceptable final route layout solution. The diagram in *Figure 8.10* shows the routing layout for a PCB application. This type of diagram can be used directly as a photomask in PCB manufacture.

The second aspect of electronic design is known as CAEE, computer aided electronic engineering. These software tools help the designer with the initial schematic and logic of the design. Within these packages, the electronic designer works through a hierarchy gradually refining and defining the electronic design. The first stage in this phase of design is the development of the input–output model. This model defines the inputs and outputs and the logical operations required to make the transformation. The logical operations required can then be built up from a series of logic primitives. These logic primitives can then be expressed in terms of some combination of AND, OR, NAND, NOR logic gates, which give the lowest level of logical function. Most of the logical elements can be described and manipulated using CAEE packages, and complex logic

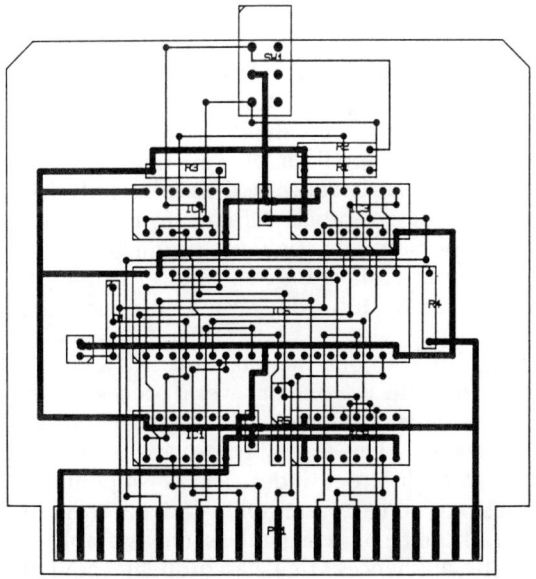

Figure 8.10 Routing diagram for the PCB

circuits can now be built up rapidly from the combination of large numbers of logic primitives. CAEE packages remove much of the tiresome manipulation required in this design process; they allow the designer to store the design of complex chips and facilitate their inclusion in application circuits.

Simulation forms an important function in these packages allowing the designer to check the design at each stage. Such a facility reduces the need to build breadboard prototypes of designs for checking purposes. Within these packages, circuit performance can be simulated to check that the logic is correct and the timing and performance under fault conditions is acceptable. CAEE packages are now integrated with CAED packages to complete the whole electronic design procedure.

Design for economic manufacture

It is an old but sadly true analogy that the transfer of a design from the design office to manufacture can be described as throwing the information over a wall and hoping that the other side can do something with it. Those of an optimistic disposition saw the introduction of CADCAM as a technology that would break through the wall and allow or even force the design and manufacture function to communicate. Although the CAM aspects of this integrating technology still lag behind the more direct design aspects, there can be little doubt that integrated manufacture has arrived. For example, certain aspects of CAD and the use of NC machine tools are now well integrated and there is a widespread pool of experience in this technology.

An important element in the chain of communication between the designer and manufacture is the production methods engineer. In an ideal world, the designer would produce a design that not only met the technical and esthetic requirements to satisfy the customer but would also represent the optimal design as far as economic manufacture was concerned. In the past, this compromise between specification and cost was approached in a somewhat haphazard fashion with the production methods engineer bearing the brunt of the work. In these more enlightened days, it is seen that to a large extent cost savings at the shop floor level are limited and the most significant decisions as far as cost is concerned have already been made before the design leaves the design office. Of course, in industries such as automobiles, computers and domestic appliance production, it has been essential to maintain a rigorous control on costs and in many ways these industries have led the way towards design for economic manufacture.

It is convenient at this point to consider manufacture in two types: firstly, component manufacture that may use one or several processes from the vast range of different processes and, secondly, the assembly of these components to form a final product. The assembly process is in some respects quite similar across such different industries as shipbuilding, aircraft manufacture or electronic board assembly. Indeed it is the recognition that ships and aircraft can be built using the same assembly line techniques as cars and domestic appliances, that has brought about the increases in productivity in both ship and aircraft manufacture.

Component design

Some of the common manufacturing processes, such as welding, forging, machining and casting, have been in the process of improvement since the industrial revolution, and may have now reached such a state that further development would be relatively unrewarding. Thus further progress must be made by ensuring that the designs exploit the processes to the utmost. Consequently, the topic of design for economic manufacture is attracting increased attention. So far the development of this subject has been somewhat piecemeal, with various attempts to define rules and techniques for efficient design from a manufacturing point of view. Machining processes, and consequently design for machining, are probably the most computer integrated aspects and will serve as a good example.

Design for economic machining
There are two levels of consideration for design for economic machining. Firstly, there are general guidelines and rules of thumb that are based upon the ease with which a particular feature of a component surface can be created. For example, such guidelines would indicate that square bottomed blind holes are to be avoided if at all possible because they would require a complex, slow, and hence expensive production process. A second and more detailed level of consideration occurs once the basic manufacturing process has been selected and there is a need to minimize the cost of that particular process for a particular product.

In considering the economics of machining it is usual to subdivide components, initially into different general shape groups such as rotational parts and nonrotational parts. It is then necessary to further divide components using, for example, aspect ratio values based upon length/diameter and the like.

The first batch of guidelines can be very general and relate directly to the optimization of material cost and minimization of material removal. They can be listed as follows:

- design to use standard components,
- minimize machining by use of preshaped workpieces,
- use preshaped workpieces designed for a previous similar job,
- attempt to standardize on the machined features,
- choose work material that will result in minimum total cost,
- check for compatibility with standard sizes from suppliers of raw material and design components to minimize machining.

Consideration of these guide lines indicates the importance of material cost, minimizing machining times, and minimizing tool set-up and change times.

A further set of guidelines is related to the relative motion of the tool to the workpiece and problems of collision of tool, workpiece, tool holder, and work holder:

- avoid features that are impossible to machine, for example the features shown in *Figure 8.11*,
- avoid features that are difficult to machine and will require special tools or fixtures,

- minimize features that are expensive to produce even though standard tools are used,
- avoid machining components all over,
- ensure that tool and tool holder do not interfere with the workpiece or work holder using standard tools and tool holders. *Figure 8.12* shows a typical example.

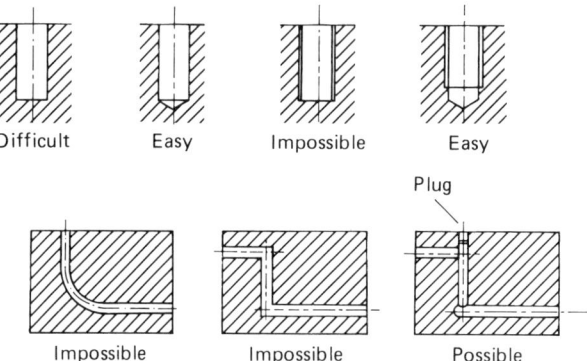

Figure 8.11 Features that are impossible to machine (from *Fundamentals of Metal Machining and Machine Tools*, by G. Boothroyd, McGraw-Hill)

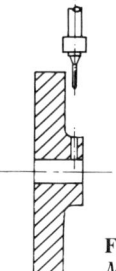

Figure 8.12 Interference of tool holder and workpiece (from *Fundamentals of Metal Machining and Machine Tools*, by G. Boothroyd, McGraw-Hill)

For rotational components, it is useful to categorize their shape into length/diameter (L/D) aspect ratios greater than 0.5, in the region 0.5 to 3 and above 3. The major factor in using such a classification is to group components on the basis of the most appropriate work holding technique. Selection is based on convenient access of the tool to the surface to be machined without the need to remove the workpiece for reclamping in a different position.

Components with an aspect ratio less than 0.5 are usually held on the turning machine by means of a face plate. Such a means of work holding naturally leads to the following guidelines for the component designer:

- design the component so that machining is not required on the unexposed faces when gripped on the faceplate,
- the diameters of external features should increase gradually from the exposed face and internal diameters should decrease from the exposed surface, see *Figure 8.13*,

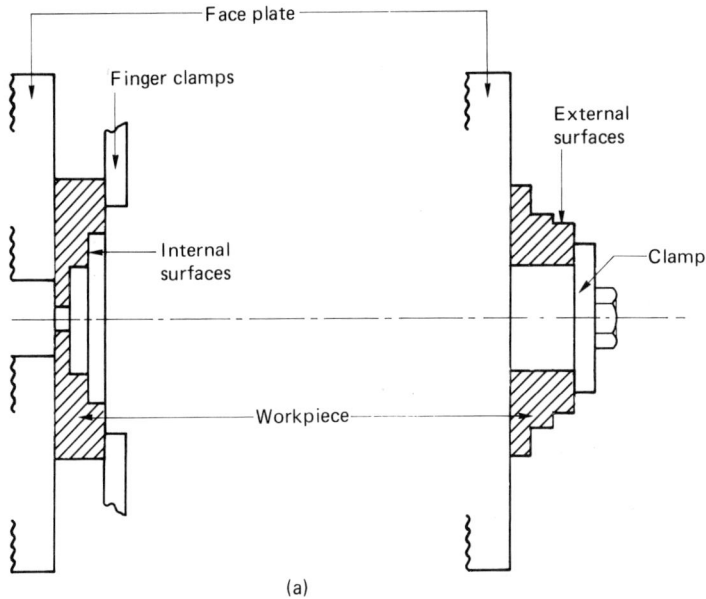

(a)

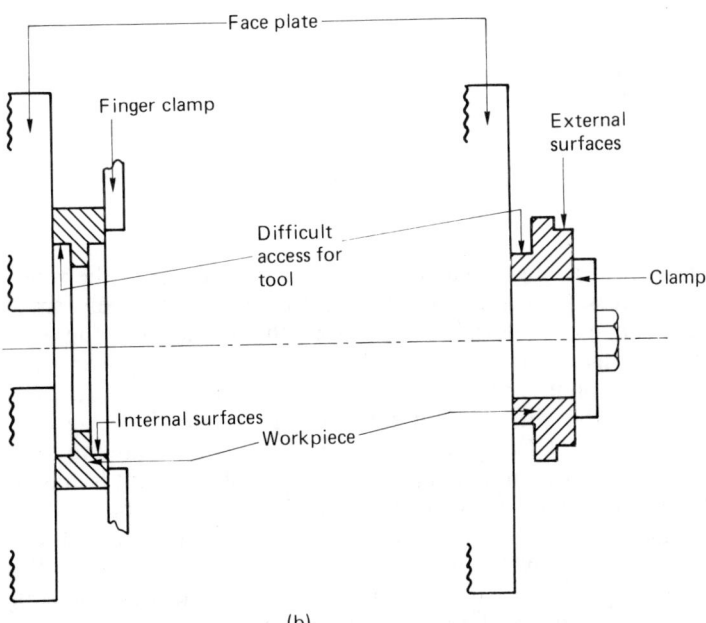

(b)

Figure 8.13 Internal and external surfaces on a surface plate-mounted workpiece: (*a*) recommended arrangement, all machining carried out on a single setting, (*b*) non-recommended arrangement, machining requires the settings of part on faceplate

- if possible make all machined surfaces concentric cylinders or plane surfaces normal to the axis. Short tapers or chamfers are not particularly difficult,
- avoid auxiliary holes inclined to the axis of the workpiece.

Components with aspect ratios between 0.5 and 3 will usually be held in a chuck and once again it is the accessibility of the tool that determines the guide lines:

- external diameters should increase towards the chuck and internal diameters decrease towards the chuck,
- there should be no recesses or grooves on parting off surfaces, as illustrated in *Figure 8.14*.

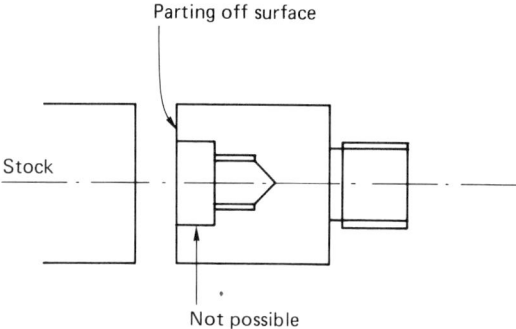

Figure 8.14 Grooves on a parting off surface

Long workpieces with aspect ratios greater than 3 will usually be supported between centres and may require a steady to reduce workpiece deflection under cutting forces. Both guidelines relate directly to these two aspects:

- ensure that the workpiece is sufficiently rigid to withstand the machining forces,
- avoid internal surfaces.

This brief review of a single machining process selected from the range of machining processes (which are themselves only a relatively small subgroup of the range of material processing methods) indicates the level of detailed knowledge of part, tool and process that is required. The magnitude of the problem and the lack of understanding of the processes lies at the root of the difficulty of introducing systematic design for manufacture as part of a CIM system. Computer systems are able to offer database structures to store and manipulate this type of data, but it may be some time before practical design aids are available on a wide basis. Even so there are signs that design for manufacture is infiltrating the design process much more extensively than in pre-CIM days. The reason for this evolution lies strangely not with the increased design power available through CAD but because of the inflexibility of shop floor automation.

The highest level at which design for manufacture is influenced by CIM is at the business level. In a CIM-based business, the facilities available on the shop floor will be a reflection of the business aims of the company.

When automation is introduced constraints invariably arise in comparison with a manually controlled and operated system. For example, automatic machining facilities consisting of NC machine tools in conjunction with automatic part and tool handling will have been installed to deal with the manufacture of a family of parts. The existence of this type of facility gives the company an efficient means of producing such parts and naturally management will seek to exploit that advantage to the full. The design office will further this business aim by designing parts that will match the production facility. Even with the 'flexible' manufacturing processes, the cost penalty of a design taking manufacture outside the capability of the production system is now much greater than with older manually based systems. Two examples will aid understanding. Many machining systems are based upon a workholder pallet system. This technique eases many of the handling and workholding problems but introduces a size limitation, typically being expressed as a maximum volume for the workpiece. For example, the handling system may be able to handle all parts of a particular type provided that they fit within a 1 metre cube. Further, systems can often take a number of components on a pallet provided they can be arranged within the limiting volume. Under these restrictions, the designer will seek to meet requirements by building up larger structures from component parts that lie within the volume limitation of the system. In a similar manner, each machine tool will have a restricted range of tools available in its magazine. These tools can be used automatically by the machine without the intervention of an operator or a tool transportation system. It is clearly more efficient to restrict the choice of machined features available to the designer to those that are available from a single magazine of tools. Many parts will require not only the use of a large number of tools but the use of several different machine tools. Efficient use of modern manufacturing systems requires the designer to make careful use of the standard tools and processes, and it discourages the introduction of nonstandard features or the use of a tool range that cannot be contained by a single tool set.

On the shop floor, automation is imposing constraining influences and in response design guidelines are evolving. For example, many machine tools and robotic systems are programmed for particular component manufacture using CNC machines. Programs generated in this way use software and hardware that is the actual manufacturing facility. Thus it is the use of the facility itself for programming that forms the ultimate constraint. Increased interaction between the designer and shop floor will allow the creation of a design that not only meets the designers requirement but which has also been introduced to the shop floor in the form of a piece part program generated by machines on the shop floor and tested in that environment.

Computer-aided process planning

Process planning is the process of defining the sequence of operations that a part has to pass through during its manufacture. Traditionally this sequence is generated manually and appears on a route sheet that identifies both the processes and the machines to be used. In some instances the process information is little more than an instruction to manufacture the part but at a more detailed level the route sheet may also describe the conditions for the process such as the speed and feed of cutting in a machining process or the arc current and voltage in a welding process.

There are a number of disadvantages of manual process planning, the major ones being caused by inconsistency. For example, it is not unusual for a process planning engineer to specify different routes for a family of parts when in fact a single common route would be suitable. Also different planners may suggest different routes for the same part, each expressing their own preference. Further there is no way of being sure that any route is optimal.

In response there has been an interest in the automation of the process planning process within CIM, in the form of the introduction of Computer Aided Process Planning (CAPP) software packages. Automatic Process Planning has two forms: Retrieval type CAPP and Generative type CAPP.

In retrieval type CAPP, parts are classified into family groups such that each group has a standard process plan. These process plans are stored as files within the computer and can be called up whenever they are required. For new parts that are related to the original family the process plan may need to be modified or varied and hence this form of CAPP is sometimes referred to as a 'varient system'. *Figure 8.15* shows the information flow in

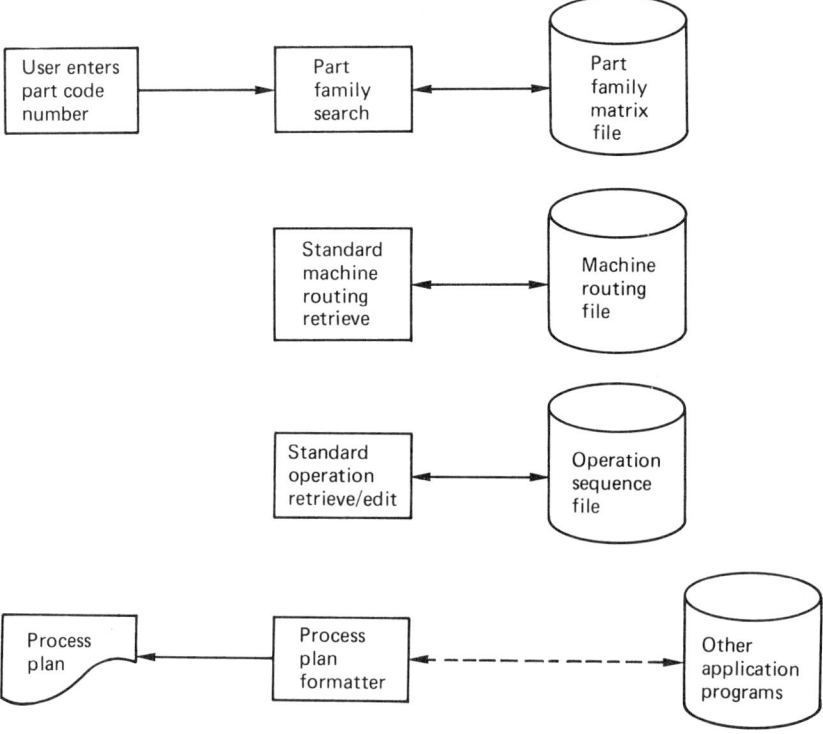

Figure 8.15 Information flow in a retrieval-type CAAP system (from *CAD/CAM*, by M. P. Groover and E. W. Zimmers, Prentice-Hall)

```
. . . . . . . . . . . . . . . . . . . . . . . . . . . . . . . . . . . . . . . . . . . . . . . . . .

PART: 640/2549 L/O: CONVENTIONAL        POS: 1-5

. . . . . . . . . . . . . . . . . . . . . . . . . . . . . . . . . . . . . . . . . . . . . . . . . .

          COST CODE        LOCATION
          OPERATOR INSTRUCTIONS        TOOLING        SET    RUN

      1   203              CAPSTAN LATHE              20    3.072
    - 1   FEED TO STOP                                      0.150
    - 2   ROUGH TURN 25.5 X 30.5 LENGTH                     0.459
    - 3   FINISH TURN 24.90 X 30.5 LENGTH                   0.253
    - 4   DRILL 15.0 HOLE X 67 DEPTH                        2.010
    - 5   PART OFF 59 MM LONG                               0.200


. . . . . . . . . . . . . . . . . . . . . . . . . . . . . . . . . . . . . . . . . . . . . . . . . .

OPTION                    - 15 entries in menu
KEY      ABBREV           OPERATOR INSTRUCTIONS

     1   CLDC             LOAD TO COLLET
     2   FDST             FEED TO STOP
     3   RFFN             ROUGH AND FINISH TURN __ MM DIA X __ MM LONG
     4   THRD             TURN THREAD DIA __ MM X __ MM LONG

. . . . . . . . . . . . . . . . . . . . . . . . . . . . . . . . . . . . . . . . . . . . . . . . . .

* SW SCH

*

. . . . . . . . . . . . . . . . . . . . . . . . . . . . . . . . . . . . . . . . . . . . . . . . . .
```

Figure 8.16 Route sheet generated by C plan (from *C plan Computer-aided Process Planning*, CAD Centre, Cambridge, UK)

a retrieval-type CAPP system and *Figure 8.16* shows a typical CAPP route sheet generated by the CAPP system C-plan.

Generative process planning is much more difficult to achieve and as yet remains very much developmental. However some generative process planners do exist for limited process application areas. In generative process planning, the computer creates the process plan using a set of algorithms, based upon general rules about manufacturing routes, without reference to any previous specific plan. The input to such a system would have to be a detailed description of the part to be manufactured including a geometrical description and material specification.

Although CAPP has not yet achieved the same degree of application as CAD for component part design the advantages of this aspect of CIM are well recognized: process rationalization, increased productivity from process planning staff, reduced planning time and hence lead time for manufacture, improved communications between design and shop floor, and improved process specification.

Although the CAPP packages that are available offer flexible, interactive environments for storing, manipulating, and retrieving production process data the task of inputting that data in the first instance is formidable. The family of machining processes is the most well understood and the most widely found application. There are an increasing number of cases where companies with a limited range of other well understood processes have used CAPP effectively.

Design for assembly

One area of manufacture where a good tool for computer-based design for manufacture exists is mechanical assembly. Work by G. Boothroyd has found wide recognition and forms the basis of a software package that aids the designer to improve the assemblability of a product. The package consists of various levels of detail from general rules that give overall advice on design down to detailed analysis of costs of feeding individual components.

Although some considerations, such as reduction in the number of parts in an assembly, will have advantages for manual assembly, automatic assembly, and robotic assembly, other topics, for example the feedability of parts, will be directly related to specific assembly methods.

Boothroyd's approach starts by identifying the most appropriate assembly method from: manual assembly, robotic assembly, automatic assembly. The criteria used for such a decision are annual production volume, number of parts in the assembly, anticipated market life of the product, number of product styles and frequency of design changes.

The extreme ends of the spectrum are easily identified; for mass production (>700 000/year) of an assembly with a few parts (less than 7) and where there are few design changes envisaged over a three year market life, the recommendation is to use automatic assembly. In contrast, for low volume production of an assembly with a large number of parts where the design is not stable and has variants, manual assembly will be the most economic.

Once a general method of assembly has been decided, a more detailed analysis can be undertaken. If desired the designer can return to the general method selection and repeat the procedure based upon alternative methods.

For both automatic and manual assembly, major cost savings can often be achieved by a reduction in the number of parts in an assembly. The analysis assesses this situation by asking three questions:

- Do adjacent parts have to be made from different materials?
- Do the parts move relative to each other?
- Do the parts have to be separated to allow assembly, disassembly, or servicing of other parts?

If the answer to all of these questions is negative then the analysis suggests that the design is less than 100% efficient with respect to part count. However, although a particular assembly design may have the potential for combining two or more parts into a single component (based upon the above rules), the cost increase may be prohibitive and it may be more economic overall to keep the parts separate.

For manual assembly, the difficulty of assembly is based upon difficulty levels for manual handling and for manual insertion. Factors that affect handling difficulty are:

- Does the handling require tools?
- Are two hands required to handle and sort the parts?
- Are two or more persons or mechanical assistance required for handling?

In a similar manner, the difficulty level for manual part insertion is considered. Difficulty is associated with small clearances between parts to be assembled, the provision of chamfers, the requirement to hold down parts whilst adding further parts, obstructed access and restricted vision. Further difficulty is added by the requirement to use assembly tools such as screwdrivers.

Analysis for automatic assembly is more complex and detailed. It is based upon a coding system similar to that used for group technology. Levels of difficulty are assessed for the automatic feeding and orientation of parts and for automatic insertion.

Each part has an assemblability code and each code an associated cost factor. Hence a comparative cost for different assembly designs can be assessed.

The code system is based upon the shape envelope of the part, which is degenerated into a simple prismatic shape ignoring small details. Essentially, this information identifies types of feeder and feed rates. Other code digits are then used to identify distinguishing geometric shape features and symmetry. Overriding codes are used to deal with parts that are known to have characteristics that produce feeding problems such as nesting, and light, delicate, sticky, and flexible parts.

As with manual assembly, the difficulty of part insertion is associated with the clearance between parts to be assembled and the provision of chamfers or lead-in surfaces. Self-locating and self-securing joints are an advantage; in contrast the use of fasteners such as screws, rivets, and spring clips carries cost loading.

A manual analysis based upon Boothroyd's is possible using the *Design for Assembly Handbook*. This is available from the Industrial Centre Ltd, University of Salford, UK. A software package exists that removes much of the tedious calculation enabling quite complex assemblies to be analyzed quickly and efficiently. The package is available from the Department of Manufacturing, University of Rhode Island, USA.

This package follows the line of the analysis above and leads the designer through the analysis via a series of menus and question and answer sessions. The package allows the designer to modify the data within the package so that it matches the conditions within a particular assembly shop by specifying the operator and supervisor costs, overheads, shift working and operator efficiency. Similarly, for automatic assembly the user can input current costs for typical feeders, workheads and indexing tables.

Group technology and coding systems

Group technology is a manufacturing philosophy that suggests that there are advantages in identifying and grouping parts into part families. By manufacturing individual parts of different detailed design in family groupings, it is suggested that set-up times can be reduced and process plans can be standardized. As a consequence of these increases in manufacturing efficiency, there will also be corresponding advantages in terms of better scheduling, reduced inventory and better tool utilization.

GT is now recognized to offer many attractive advantages particularly in conjunction with automatic handling. The task of achieving an appropriate and valid grouping of parts can be a formidable difficulty. This task is usually approached in one of three ways. The least sophisticated but also least reliable method is by visual inspection of the parts, arranging them in groups based upon their physical appearance. Basic design attributes that can be used directly are external shape, internal shape, length to diameter aspect ratio, major or minor dimensions etc. More indirectly, characteristics based upon known but not always visually determined features could be material type, part function, tolerances and surface finish.

An alternative technique is to analyze the process route for the parts and to group parts that employ the same tool grouping. This approach can have advantages in preparation for computer aided process planning but suffers from the disadvantage that it is based upon accepting the validity of the existing process sheets. The final and most systematic approach to be discussed is the use of some method of classification or coding. This technique is the most time consuming to implement but also has the potential for being the most powerful, especially within computer integrated manufacture.

Over a number of years many classification and coding systems have been developed and proposed. Several of these systems have been made commercially available but so far no universally acceptable all-embracing system has been produced and experience shows that within all systems the application company has to customize the scheme to meet their particular requirements.

The growing importance of part classification and coding within CIM makes it appropriate to give some attention to a review of some individual schemes and give a brief description of their basic structure.

A commonly found manner of identifying parts within a system is by the use of part numbers. In this type of system, each part is given an individual number as it is designed. No attempt is made to relate the features of the part to the number and as a consequence the part number does little more than indicate the time when the item was designed or the order of a series of revisions to a design. In some instances further information may be added; for example, a code letter attached to the number may indicate the department or factory where the part was designed or made, or the parts destination, the assembly for which the part is a component. This way of identifying parts carries no relevant information for design or manufacture and has the deficiencies that it can easily lead to duplication of designs and does not give any indication of manufacturing route.

Other more systematic methods of part identification are much more useful and are able to describe the part with useful accuracy. In the Opitz system parts are classified on the basis of their shape and the machining processes required to produce that shape. Shapes are given code numbers indicating membership of a group of common features. The basic code consists of five digits. The first digit describes the component class in the most obvious division, that of rotational and nonrotational parts. This first digit also gives an indication of the proportions of the component shape giving a code identification for parts that would be considered as disks or long bars or cubic blocks etc. The remaining digits further identify features of the shape in a more refined and detailed manner. The diagrams in

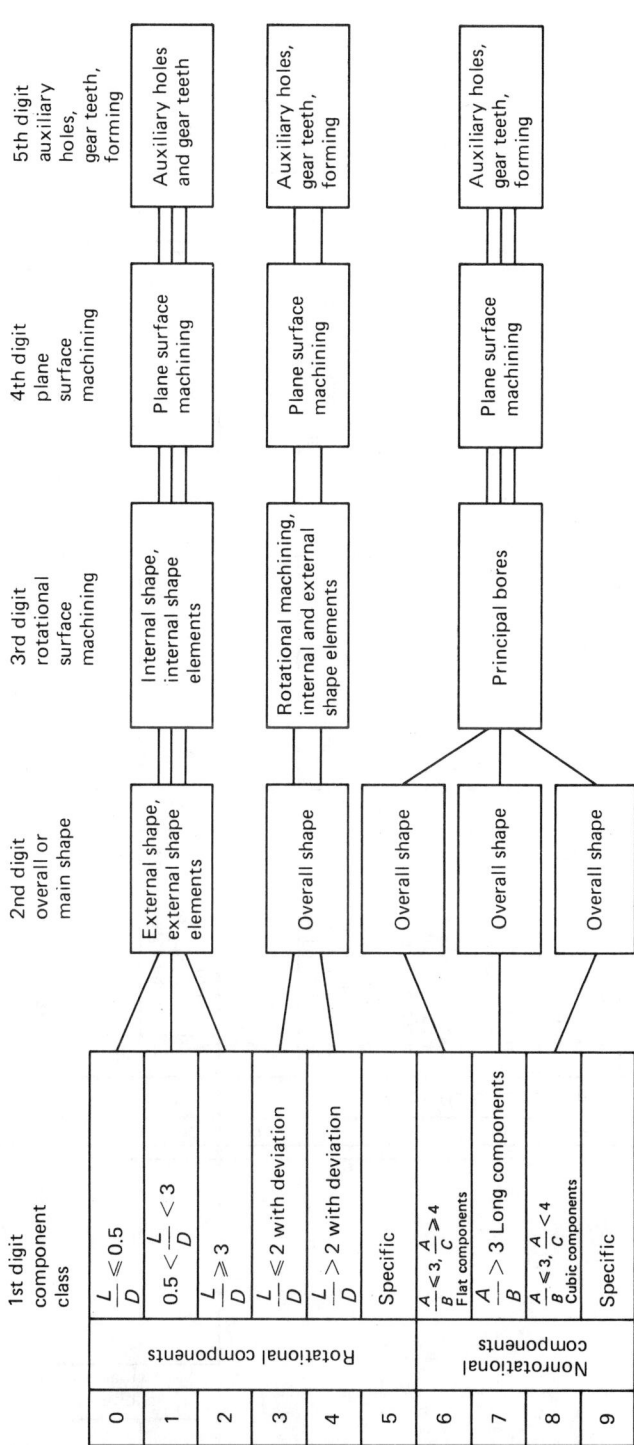

Figure 8.17 Geometric code for classification of machined components (by permission Verlag Wu. Girardet)

	1st digit	2nd digit	3rd digit	4th digit	5th digit
	Component class	Overall shape	Principal bore, rotational surface machining	Plane surface machining	Auxiliary hole(s), forming, gear teeth
0		Rectangular prism	No rotational machining or bore(s)	No surface machining	No auxiliary holes, gear teeth, and forming
1		Rectangular with deviations (right angle or triangular)	One principal bore, smooth	Functional chamfers (e.g. welding prep.)	Holes drilled in one direction only
2		Compounded of rectangular prisms	One principal bore stepped to one or both ends	One plane surface	Holes drilled in more than one direction
3		Components with a mounting or locating surface and principle bore	One principal bore with shape elements	Stepped plane surfaces	Holes drilled in one direction only
4		Components with a mounting or locating surface, principle bore with dividing surface	Two principal bores, parallel	Stepped plane surfaces at right angles, inclined and/or opposite	Holes drilled in more than one direction
5		Components other than 0 to 4	Several principal bores, parallel	Groove and/or slot	Formed, no auxiliary holes
6		Approximate or compounded of rectangular prisms (Not split)	Several principal bores, other than parallel	Groove and/or slot and 4	Formed, with auxiliary holes
7	Nonrotational components	Components other than 6 (Not split)	Machined annular surfaces, annular grooves	Curved surface	Gear teeth, no auxiliary hole(s)
8	Cubic components $\dfrac{A}{B} < 3,\ \dfrac{A}{C} < 4$	Approximate or compounded of rectangular prisms (Split)	$7 +$ principal bore(s)	Guide surfaces	Gear teeth, with auxiliary hole(s)
9		Components other than 8 (Split)	Others	Others	Others

Notes on 2nd digit grouping: values 0–5 are Block and blocklike components; values 6–9 are Box and boxlike components (6, 7 = Not split; 8, 9 = Split).

Notes on 5th digit grouping: values 1, 2 = No gear teeth, no forming; values 3, 4 = Related by a drilling pattern; values 5, 6 = Forming, no gear teeth; values 7, 8 = Gear teeth, no forming.

Figures 8.17 and *8.18* show the initial breakdown of the classification system for machined components and an example of the system applied to cubic components. Clearly this system is dedicated to machined components. Although the principle of the system could be applied to nonmachined components, it would require the definition of another classification system.

The Brisch system overcomes this lack of flexibility by including the definition of the fields of classification as a part of the component code number itself. Codes that relate to a fixed field type of classification are known as polycodes and the Brisch system is defined as a monocode. Within the monocode system, the code digits are read so that each digit indicates the field within which the next digit operates. This flexibility gives the coding system considerable power but requires more skill in the development of a particular application. As with the Opitz system, the code appears as a series of digits and once a field has been identified by the first digit of a pair the second digit operates in a similar manner to Opitz with the field being subdivided into generic groups.

Robotic assembly

Although design for robotic assembly shares most of the characteristics of design for assembly in general, and design for automatic assembly in particular, there are several interesting differences. Robotic assembly occupies an application window somewhere between manual assembly and automatic assembly using dedicated assembly machines. In this position, robotic assembly is often a compromise being slower and more expensive than hard automation for large batches and more expensive and slower to bring into production than manual assembly for small batches. Not very many years ago many observers were pessimistic that robotic assembly would find a viable place against the well-developed competition. However, more recently, assembly by robots has established itself and generic robot-based assembly cells are increasing in numbers.

This gradual change presents the designer with increased possibilities. Firstly, it becomes increasingly important that the designer produces designs that fall within the generic capability of the robot cell. The designer is no longer constrained to the very narrow variations in design allowed by hard automation, nor does he have the freedom to assume that the operator on the shop floor will work out how to carry out the assembly by some means or another. This has the effect that the designer must seek a design that is as simple to assemble as that demanded by hard automation in terms of part feeding and insertion. However the robot can carry out much more manipulations of parts during assembly and so there is less of a priority for the assembly to be one sided or for the insertion to be a single linear movement.

The major freedom given to the designer by the introduction of robotic assembly is not in the design of the device itself but in the ability to design families of parts or to modify designs such that manufacture can still be carried out within a single generic cell or system. Although designing for the batch of one still has some way to go, the target is still on the horizon and the designer's freedom and power to change designs without changing the manufacturing route or tools is increasing steadily. The knack here is to produce designs that appear to the manufacturing system tools to be identical, i.e. they present identical gripping and location surfaces to the handling and fixturing tools. The use of encapsulation in the manufacture of turbine blades is a non-assembly example of this philosophy. The encapsulation of the blades, which can be of different complex shapes, makes the different blades appear to be identical as far as handling and fixturing are concerned. In the case of assembly, it may be possible for the designer to introduce a common feature; for example, a hole that can be used as a common location or gripping point across a range of quite different parts.

Process innovation

Most manufacturers and designers tend to be conservative and reactionary in their approach, only exploring or exploiting new and novel processes where the new process offers a facility not available by more traditional processes of which they have experience.

As the demands upon manufacturing processes has increased, resulting from the use of more difficult materials, higher tolerances, and more complex shapes, new manufacturing processes have been introduced. At first these processes were regarded as novel but, as they have established their application niche, they are no longer regarded as anything out of the ordinary. Processes such as electrical discharge machining, chemical machining, electron beam welding, explosive welding, creep feed grinding and powder forming have all passed through this route. It could be considered that many laser-based processing methods, liquid jet machining, single crystal casting proceses and many others are still in the process.

In addition to process innovation based upon technical advances, economic and social situations can also have a surprisingly significant effect upon process development and utilization. For example, the 1960s and 1970s was a period of considerable social and economic change. During that time, and recently, there has been an increasing awareness of the environment and the polluting effect industry can have. This concern generated legislation that in turn changed the practices in many industries and has not only lead to cleaner, quieter processes it has also paved the way for the introduction of automation and the humanization of the workplace. An additional influence was the oil crisis that took place during that period. This situation led to a more general awareness of the finite nature of the world's resources and resulted in an interest in and application of energy and material conserving processes.

These technical, economic and social influences are discussed here because it is important to stress the dynamic nature of manufacture and the necessity for the designer to be aware of this situation. Over the rest of this century, the pace of change in material processing and design will increase and it will be interesting to observe and participate in a process whereby manufacturing technology translates into reality the ideas of designers and the wishes of customers. Corresponding changes will have to occur in planning and control systems; these are considered in the next chapter.

Further reading

GROOVER, M. P. and ZIMMERS, E. W. (1984) *CAD/CAM: Computer-aided Design and Manufacture,* Prentice-Hall International Editions

ROONEY, J. and STEADMAN, P. (1987) *Computer-aided Design,* Pitman/Open University

McKISSICK, M. L. (1987) *Computer-aided Drafting and Design,* Prentice-Hall Inc.

MEDLAND, A. J. and BURNETT, P. (1986) *CADCAM in Practice,* Kogan Page

BOOTHROYD, G. (1981) *Fundamentals of Metal Machining,* McGraw-Hill International Student Edition

BOOTHROYD, G. *et al.,* (1983) *Automatic Assembly,* Marcel Dekker

APPLETON, E. and WILLIAMS, D. J. (1987) *Industrial Robot Applications,* Open University Press Robotics Series

Chapter 9

Planning and control in a CIM environment

If the production plan can be improved so that a customer order is saved or a delivery postponed, then it must make sense to have the computers work longer hours. If the 'overtime' is during the night, when the CAD and other office terminals are not being used, the only extra cost is the electricity. The benefit can show immediately on the bottom line.

Planning is the process of organizing material and component availability and of optimizing the use of productive capacity in a manufacturing organization. Control is the process of making the plan happen, including monitoring of production progress.

Many of the well-publicized examples of successful planning and control come from large companies with high volume production and world-wide markets. These techniques are continually becoming available to other industries and smaller companies.

System evolution

The wide introduction of on-line computer systems into manufacturing industry began in the late 1970s. In spite of all the prognostications that 'Our people will never understand it', 'You can't cost justify that sort of system in our industry' and 'The IS people will never keep the system up', the next ten years saw a rapid growth of on-line planning and control systems.

This growth was largely built around MRP as a planning tool, as described in Chapter 5. The growth reflected that MRP needs accurate stock data, the ability to recalculate the plan frequently and the immediate availability of the output. These requirements could be satisfied by on-line systems, so that MRP could then be widely installed in small companies as well as in the multinationals. In the same way, integration of computers is going to cause a fundamental shift in planning and control systems.

MRP has evolved into MRP II. As described earlier, MRP stands for Material Requirements Planning; MRP II stands for Manufacturing Resource Planning (the II identifies which acronym is intended). The distinction is basically that MRP II includes capacity planning; in the context of MRP II, the acronym MRP alone refers to the numeric calculation of component and material requirements. The gradual addition of capacity planning to MRP systems clearly required a new acronym; by the end of the 1980s most companies do some form of capacity planning calculation on a computer, even if only at the MPSP level with a spreadsheet package, so by definition these companies are into MRP II. It is still convenient to use MRP to refer to the number-crunching exercise required to calculate component and material requirements; capacity planning within CIM is discussed in more detail below.

International competition in manufacturing industry is still increasing, so that the companies that turned to MRP for improved planning and control will now require the improvements in productivity that CIM can provide, based on the components of CIM architecture described in Chapter 7. Increased automation on the shop floor, driven by the same competitive pressure and by the need for guaranteed quality, will predicate tighter planning and control. Flexible machining and assembly cells are very expensive, often requiring a much higher investment than computers for a new planning system. Major investments in capital equipment make little sense if the output cannot be properly incorporated into a company-wide planning system.

MRP and MRP II systems depend on human beings to interpret the output of the computer and to make the decisions that require judgment (as opposed to calculation). In future, there will be fewer people on the shop floor to take these decisions, so CIM planning systems will be unable to rely to the same extent on human assistance in finalizing the details of the plan. In any case, rapid changes in the market place will reduce the relevance of past experience, which humans are currently much better at assimilating than computers.

Hence, more function will be required in planning systems, based on new application software packages and extensive use of database technology. As the systems become more complex, deeper understanding of them, and considerable training, will be required. Using a production control system to run a CIM factory may become comparable to flying a modern passenger airliner, requiring aptitude, training and continual testing of competence; simulation may be an effective means of teaching these skills. These CIM planning systems will need to evolve from existing production control techniques.

Material control within CIM

MRP (i.e. the calculation of component and material availability) will still one of the main engines of CIM planning and control systems, adapted to the architecture described in Chapter 7. MRP could run either on a mainframe computer or on a PC that is part of the shop floor network, the choice depending on the CPU power and disk space required. Data about stock movements could come from VDUs connected direct to the mainframe or via the shop floor network. This flexibility can therefore conform with a CIM architecture; MRP applications can be moved according to the users requirements.

However, MRP will become a small, almost invisible, part of the total system, leading incidentally to occasional discussion as to whether MRP is obseleted by techniques such as JIT (see below). MRP will still be there, but other aspects will become more important.

Bills of material within CIM

As discussed in Chapter 5, there are many possible different versions of a bill of material. For instance, the version prepared by the designer and used for costing calculations, will be of little interest to people in assembly, who need to see those components allocated to each stage of the build process. Different bills of material are often required when selling into different countries, since electrical regulations and the more mundane matter of labels and instructions in different languages can require different versions of the finished product. Rather more subtly, functionally identical subassemblies may contain different components; for example the only difference between two subassemblies could be the brand of electric motor included, a distinction that is only important to customers in countries where just one of the brands of electric motor is supported for spare parts and servicing. Service departments often require extra data on the bill of material; for instance, components of purchased assemblies may need to be stocked and controlled.

The number of levels in the bill of material needs to be related to control procedures on the shop floor. In general, the more levels and the more subassemblies, the tighter can be control of material movements; a balance must be struck between the need for control and the cost of data collection. As design and manufacturing systems become integrated, decisions on how the bills of material should be structured will need to be incorporated into the engineering change control system.

Some companies include the finished product in the bill of material database, e.g. item number 456 is the two metre model with air conditioning and three legs, item 457 is identical except it has no air conditioning. Another approach is to define a base product and separate item numbers for each option or feature (modular bills of material). Among the complexities introduced by this approach is the use of negative quantities in the bill of material to allow for combinations of options and features. A deciding factor between these two approaches will be how large a range of extras the company wishes to offer, and how customer orders are to be processed.

Customer orders may well be entered by computer-to-computer link from the customer's computer, subject to some careful attention to security and credit worthiness. Each customer order will need to be netted against the master production schedule, as discussed in Chapter 5. This process may not be straightforward, particularly if the MPS includes percentage probabilities for subassemblies. Materials and components will be allocated, and new orders sent in turn to suppliers, all within a few minutes, and untouched by human hand. This communication between customer and supplier is described further in Chapter 11, including the exchange of CAD and text data.

Structuring the bill of material is, of course, a well-explored subject; business-based solutions will be needed for each manufactured product if CIM systems are to process the data automatically and take decisions. In outline, the solution will be that the manufacturing requirement decides the basic format of the bill of material, and that the database facilities will be used to give design, costing and service departments the views which they need. However they are set up, the bills of material will still need to integrate with the logic of the MRP system.

MRP logic

Some significant extensions to conventional MRP logic will be required if the order entry is to be fully automated. For example, it may take some time to investigate whether a short lead time customer order can be accepted, so that components would need to be held in a semiallocated condition, to avoid them being used for another customer order while enquiries are proceeding about the first order. However, the later order may be on normal lead time, more profitable or for a more important customer; maybe, if all the components are available for the later order, it should be accepted and the earlier order refused.

Also MRP will need to become more pro-active. Few MRP systems are geared to answer the question, 'What can be built with the components and material in stock?' The customer would be told that it would take three weeks to deliver the 3 metre optional support bar, without the computer having the 'intelligence' to ask if the 3.5 metre bar, which happened to be in stock, would do instead. This logic could be added as an extension to most MRP systems, provided that data about possible alternatives were made available. It is often not easy to define these alternatives, since they would be so dependent on the individual customer situation. Perhaps, for this particular customer, the 3.5 metre bar would be no use, but a 2 metre bar in two weeks would be perfectly acceptable. This is another situation where an AI approach, as discussed in Chapter 7, would seem suitable to build up a history of possible alternative options as a guide for future decisions.

There are other issues, including MRP 'nervousness' — the system will have to be sensible enough not to send half a dozen messages a week to a supplier changing the quantity of washers to be delivered in six months time — even though the requirement may indeed have changed that frequently as customer orders are accepted or modified. This problem can be tackled by setting up a matrix relating two factors: (a) percentage change in order size; and (b) timescale as a percentage of lead time. For example, if MRP calculates a 10% change in quantity of an order due in six weeks time, this being twice the supplier's lead time for the item in question, then the CIM system should decide to wait a while before passing this change onto the supplier. A 50% change would perhaps be passed on immediately, to give the supplier maximum time to react.

Maintaining inventory accuracy
It is familiar ground that, as systems become more automated and real time, they are increasingly dependent on accurate data. Education of all the people responsible for material movements is a prime requirement. Perpetual inventory, or cycle counting, offers an important technique in maintaining accurate stock figures; numeric targets for inventory accuracy

should be set, and individual responsibility assigned. An audit trail must be maintained, and an ABC (or Pareto) approach to focus on the high value items will be useful.

Different types of manufacturing industry have different requirements for inventory recording and accuracy. The need for batch recording is common in the aerospace industry, for example; if a forging fails in operation, it must be possible to trace all the other forgings which were in the same manufacturing batch, and identify which aircraft they are in. A different requirement often arises on Government contract work, when each component is to be individually identified on the purchase order, and then traced through production to the final product, a recording process that can easily cost more than the purchase price of the component concerned.

Pharmaceutical manufacture often requires that an audit be kept of the usage of each batch of raw material, i.e. it must be accurately established that all of the expensive or dangerous materials drawn from stock were actually used in products. This requirement can exceed maximum field lengths both on the database and the VDU display areas; it is necessary to maintain four or five decimal places for the weight of material component per pill and also, to unit accuracy, the millions of pills that might be in stock after a production run. The use of multiple units of measure may not be acceptable, since an opportunity for further error is thereby introduced.

Batch sizes and lead times
Determination of batch sizes and lead times within MRP will similarly become a more important exercise if CIM systems are going to implement decisions, rather than submit the difficult issues to humans for final consideration. The batch-size problem can be illustrated by an example of the use of MRP in a multiplant situation.

The problem is to give the rules to MRP that will enable it to allocate work between plants when more than one plant can make an item. For example, if 500 of subassembly X are needed next week, suppose that Plant A can make 400 subassemblies per week and Plant B 300 per week. An algorithm is needed so that MRP (without human intervention as systems become more automated) can make the best decision about the sharing work between Plants A and B. In one particular case (a set of numbers that may never be repeated), the human would perhaps spot that the existing stock (of say 100) is all at Plant C, which is the other side of the country to Plants A and B. It would actually be better, therefore, to make 600 next week, and leave Plant C's stock to satisfy its demand for the following week. But if the stock at Plant C were 300 (a full lorry load, it just so happens) and if Plant C would not be using the stock for another 3 weeks, then the decision would be different. This sort of example can be extended indefinitely with all the the various eventualities that might affect the decision as to how many to make in Plant A and how many in Plant B. The 'If . . . then' rules (if there is a full lorry load, and if there is no demand for the next three weeks, then ship from Plant C to Plant B) being built up continuously, with the ability to backtrack and explain the logic behind decisions, would again be suitable for an AI/rule-based approach.

Lead times are also vital numbers in MRP calculations; conventionally, only fixed numbers or a set of simple formulae are included within MRP software. The CIM system will tackle this well-known problem by sending a query to the supplier's computer to ask what the earliest delivery is, and maybe even starting the night shift on the required item (assuming that the suppliers computer is programmed to expect such instructions). This represents an interface with capacity planning systems, and is discussed further below. However, some important techniques that are effective in reducing lead times need to be considered first.

CFM, JIT and Kanban

This is another aspect of CIM where a historical perspective is useful. Twenty years ago, the best Operational Research schools taught EOQ formulae (Economic Order Quantity, see Chapter 5) and the consequent benefits of large lot sizes as a means of reducing unit costs. Production Managers all knew the importance of running batches for as long as possible between set-ups. Even more important was the need to keep machines fully loaded, on next month's work if necessary, in order to increase overhead absorption. The accountants were in no doubt that fully depreciated machine tools were the most economical machine tools, regardless of breakdowns and maintenance costs, which were treated as a fixed overhead. It was quite clear to everyone, engineer and finance man alike, that holding high work in progress was not expensive.

Such opinions are no longer fashionable, to put it mildly. Many people would agree that the main reason for the present dedication to 'Just In Time' production (JIT) is not so much because the sums were wrong twenty years ago, but because it was possible to hide an awful lot of sloppy planning and poor quality behind the wall-to-wall inventory; this was the cost which became insupportable.

Major automotive companies have practised CFM (Continuous Flow Manufacturing) since the introduction of the assembly line — no locked stores, no batches, deliveries daily straight to the assembly line. It should be clear that CFM usually requires MRP, and is neither an alternative nor a replacement for it. A daily call-off of the exact number of components presupposes an earlier MRP calculation to give planning numbers to the suppliers, so that they know roughly what the daily requirement will be.

MRP is a 'push' system, it launches orders in anticipation of requirements, based on (usually) fixed lead times, without checking capacity and without effectively ensuring that the other components required are going to be available (although MRP will identify potential problems, and Production Control people can use pegging data to investigate them). JIT, in contrast, is a 'pull' system — components are only made (or issued from stores or called up from suppliers) when they are immediately required, and then only to the exact quantity. This approach has application in other industries than manufacturing, including for instance civil engineering, where storage space can be very limited, e.g. high rise buildings.

Like CFM, JIT usually assumes the presence of an MRP system to do the basic calculations of component requirements. Rather than being a particular technique, JIT has been portrayed as a philosophy, incorporating redesign of the process flow, set-up time reductions, low inventory, quality improvements and a general elimination of unnecessary activities. Hence JIT provides factory managers with a framework to pursue these goals. However, cultural changes have to be achieved. Suppliers may not necessarily have the same commitment, and difficulties may be found in mixed environments where JIT operation has to be combined with more conventional production.

The well publicized Kanban system has taken the JIT philosophy a logical step further. Kanban cards (cards containing item number, description, quantity etc.) are used on the shop floor as a mechanism to communicate requirements and to ensure that no material or components can be surplus to requirements, since nothing can move without the authority of a Kanban card. Batch sizes are simultaneously reduced, with an ideal batch size of one. The success of this philosophy has been most impressive. Its prerequisites include high quality of components, well educated and careful employees (it would be not untypical for a quarter of the workers in a Japanese factory to be educated to degree level in engineering), and suppliers who are geographically close and who work to the same high standards. Kanban benefits from the relative stability of high volume automotive production, where new models are only introduced after extensive planning.

An effect of Kanban is to impose precise capacity planning onto MRP; CIM will need to achieve this same effect for a much wider range of manufacturing companies.

Shop order release under CIM

Manufacturing companies in low volume production need to control orders on the shop floor by individual batch; CFM does not apply (the characteristics of different types of production were discussed in Chapter 3). In the lower volume environment, a decision needs to be taken as to when sets of components should be released from stores so that assembly of the batch (or other processing) can begin. If all the components are not physically available, this decision requires an element of judgment; there is a risk that the missing components will not arrive in time. The risk may be trivial if the supplier has large stocks and has delivered every week for the last three years, or may be difficult to assess if the question is, for example, whether porosity will be found in the final machining operation on a casting. On a large assembly, there may be a number of components short, each with different circumstances to determine the actual date and time of arrival.

By the mid 1980s many software packages for production control could create a shortage list (either on-line or printed) for a particular batch, showing due date of the next order for each component and other relevant data. This enabled a human decision to be made, which might be to release the batch and assume the components would arrive, release the batch with a smaller quantity than originally planned (requiring a re-run of the net change MRP software) or to delay the batch, perhaps after discussion with the customer. CIM will be requiring that more and more of these decisions be taken by computer; it can be seen that the accumulation of 'If . . .

then . . .' statements in an AI/rule-based program (see Chapter 7) could well achieve this.

OPT

OPT (Optimized Production Technology) is a commercial product and it would not be appropriate to make detailed comment on it. However, the book *GOAL* deserves a mention because of the astonishing achievement of its author (Goldratt) in creating a novel out of a production control problem — and what a boost for Western manufacturing industry if Michael Caine or Robert Redford were to take the lead role in the film!

The general philosophy of OPT is widely documented and relevant to CIM developments. OPT takes a pro-active approach to tackling production bottlenecks; it uses its algorithms (the details of which are commercially confidential) to identify bottlenecks and adjust the schedule to minimize their impact on production. Flow through the factory is what ultimately matters and not factory capacity utilization; it is bottlenecks that limit the flow. An hour's production lost at a bottleneck is lost for the entire factory: an hour saved at a non-bottleneck may well be of no benefit at all, and might, in fact, unnecessarily increase work in progress. Hence the focus needs to be on improving throughput of the bottlenecks, including reductions in batch sizes and quicker set-up procedures for jobs going through the bottlenecks.

Hence the well known calculation for EOQ based on holding cost and average annual demand should perhaps be somehow modified to include whether a particular batch will be going through a work centre that happens to be a bottleneck at the time! In one way, the OPT approach is comparable to the application of ABC analysis in stock control theory (discussed in Chapter 5) whereby the important items are identified for more careful control.

Focusing on bottlenecks is likely to require changes to the conventional product costing and overhead calculations, which may be a problem in some companies — and was a real threat to the career prospects of Goldratt's hero. However, the benefits to the company's cash flow from improved shipments to customers should more than compensate.

The OPT approach also demonstrates a point that is important for CIM: it is not necessary for users to know the details of the scheduling algorithms, provided they have confidence in them, and the systems are easy to use. OPT also relates MRP with capacity planning.

Capacity planning within CIM

Whether planning of capacity is a major business problem will be very dependent upon the type of company and the state of business; clearly capacity planning is much easier if the work load is declining — one reason why it has not been a big subject in the 1980s. Capacity planning is also much simplified if the factory has been laid out, after careful simulation, for a planned production level. Planning of capacity (as introduced in Chapter 5) typically takes place in three phases, which need to be reviewed within CIM systems:

- rough cut capacity planning, as part of Master Production Schedule Planning, for major resources only, usually over a six month to two year timescale,
- long-term capacity planning, perhaps from two weeks to a year, ideally including shop orders not yet released as well as forecast orders,
- short-term capacity planning (or scheduling) of released orders to produce work-to lists for the shop floor, sometimes called operation sequencing.

Master Production Schedule Planning (the first phase) is likely to include only outline capacity planning calculations, e.g. machining or assembly capacity in units of man months. At the timescale of a year or eighteen months ahead of actual production, when the discussion is about sizes of factory, capital investment and market, not much consideration is given to the day-by-day problems that production management will face in due course. Hence the detailed planning of capacity usually starts with the second of the above phases.

Capacity planning on computer is a complex subject; really successful installations are not common, and if the technique is to be properly incorporated into CIM systems, it is important to understand some of the reasons for this.

Limitations of capacity planning systems

Given sufficiently powerful computers, capacity planning systems should be able to optimize production throughout the factory, calculate exactly when every job will be finished and exactly what the load will be on each work centre, by the minute if necessary. CIM will require this, and has made the capacity planning job more difficult by allowing the inclusion of suppliers' capacities in the calculation.

However, capacity planning systems depend on MRP to calculate the required due dates for each batch. This imposes a limitation on the accuracy of their output, and is discussed below under the heading of 'Integration with MRP'. There are other limitations in capacity planning systems.

One major problem is how to make better decisions using the output of the capacity planning calculations described above. Infinite capacity histograms for scores or hundreds of work centres, showing load by day or by week over the next three to six months, are not easy to use, whether on paper or on a VDU. This is partly from the sheer volume of data, but also because many of the apparent overloads are non-existent, since the jobs won't be able to get through earlier bottlenecked work centres. Such histograms are not much help in deciding which jobs should be subcontracted to reduce the bottlenecks, or what effect each subcontracting would have on other, maybe underutilized, work centres. Subcontracting of work creates extra operations for inspection and transport, so could possibly be slower than waiting for the queue of work to be completed and could cost more.

Apart from dealing with work centre overloads, there is the question of what to do about individual jobs that are running late. It may not be clear which work centre(s) is holding the job up, whether increasing the capacity at that work centre (e.g. by overtime) would allow the job through; maybe there are other higher priority jobs also in the queue, and if one bottleneck is removed another work centre might in turn hold the job up.

Finite capacity calculations are often optimistic, because they do not show the effect of future work, i.e. work not yet released to the factory. For example, if there is an average of six weeks work in the machine shop, a detailed finite capacity plan is likely to show that, although some jobs may be a bit late, there is no real cause for worry because every single job will be completed in about six weeks time. But in reality more work will be released to the shop within the next six weeks, so this optimistic view is quite misplaced; detailed calculation of present load needs therefore to be combined with simulation of future load. One method of tackling this problem is to extrapolate queue lengths at machines on a statistical basis, and use these values, rather than actual load, to forecast when jobs will be completed. Capacity planning calculations can be seen therefore to require some complex logic.

Logic required in capacity planning under CIM

The logic for detailed finite capacity planning (i.e. calculations based on actual capacity) must include the ability to summarize the various priority factors such as lateness on due date, important customer, accumulated cost, into a single numeric value so that queues can be sequenced. In addition, a number of other process routines are likely to be necessary:

- reduction of standard inter-operation (or move) time for urgent jobs,
- overlapping of jobs across different work centres, e.g. the first items in a batch being heat treated while the last items are still being machined,
- splitting of batches across identical machines,
- use of alternative routing data, i.e. there may be different ways of making a product that could be chosen, depending on the load at the time on different work centres (process planning was discussed in Chapter 8),
- splitting of batches to rush part of the batch through; a batch of say 50 machined items might be split, with 10 for an urgent order being sent ahead. The extra set-up cost would be accepted in order to get urgent jobs through a work centre that is bottlenecked today, but may be clear in a couple of days for the remainder of the batch to be completed. If the computers can manage it, this might be a better general solution than the trend of moving to a universal batch size of one; the EOQ calculation is still correct that large batches reduce the set-up cost per item,
- parallel operations, e.g. requiring a tool, NC program or robot program to be available at the same time as the work centre, or an operator to be available to set it up,
- sequencing job queues, e.g. by paint colour, raw material, tool required. This will avoid too-frequent tool changes or cleaning of machinery, subject to priority override for very urgent jobs. For example, it would be better to load black paint after brown paint, but if the white job were really urgent, the extra cleaning work would be accepted,

- alternative work centres, e.g. a component planned for an operation on a small lathe can (probably) be moved to a larger lathe if the small lathes are over-loaded,
- loading a facility by volume or weight, e.g. an annealing furnace,
- allowing for time dependencies, e.g. don't heat billets late on Friday afternoon for forging first thing on Monday morning,
- ability to unload a schedule if a future bottleneck is identified, e.g. if Operation 10 is planned for tomorrow, but the software calculates that Operation 80 will be held up later at a major bottleneck, then Operation 10 should be unloaded from tomorrow's schedule; another job, which does not go through the same bottleneck, could then be loaded for tomorrow. This is potentially a very important way to reduce bottlenecks. After all if computers play chess, why should they not do look-ahead calculations for expensive machine tools?
- MRP linkages (pegging, as discussed in Chapter 5) should be preserved, since it is clearly suboptimal to schedule individual items for an assembly without regard to their dependence on other items. Pegging data should be related to a database of shop orders rather than to the bill of material, since changes may have been made to a particular batch on the shop floor.

Other considerations will include logic to select jobs from queues in the most effective sequence, including for example critical ratio analysis or shortest operation scheduling.

The capacity planning software should allow for frequent rescheduling after the latest feedback, or unexpected events on the shop floor. This probably requires that some method should be available to extract a sub-set of orders from the shop order database, to avoid rescheduling the entire factory work-load each time. Feedback from the shop floor might include on-line process data giving the latest yield figures; the capacity planning algorithms could then sensibly be related to the logic in the simulation model used to design the factory layout (simulation was described in Chapter 6).

Effective management control requires a means of monitoring the production facility by analysis of job status, e.g. what percentage of jobs are running late, what percentage are actually released late to the shop, what is the value of work in progress? This data should be available from the database, using the techniques discussed in Chapter 7.

The type of planning system described above, with the priority logic, i.e. the rules, separate from the database containing details of jobs and work centres, could be regarded as an expert system. It is noteworthy that capacity planning systems with this type of feature have been available since the late 1960s.

As already mentioned, for many companies in recent years the problem has been under-utilization rather than over-utilization of capacity. If capacity limitation is not a problem, then conventional project control systems can be used for production planning. Typically applied in building projects, project control systems take activities in sequence, allow for dependencies, determine the critical path and calculate what resource is needed on an infinite planning basis, e.g. to determine how many bulldozers are needed. They may do some load smoothing for noncritical activities, but do not usually schedule activities on the basis of availability of resources as is required in a capacity planning system.

Plant maintenance
Plant maintenance will be increasingly important in automated CIM factories; indeed shortages of the skills necessary to maintain the complex electromechanical systems may be a significant limitation on the development of CIM. In the context of capacity planning, planned maintenance should be fitted by the scheduling system into gaps in production requirements, rather than done at preallocated times without regard to the load on that work centre at the time. There will also be a need to respond to breakdowns by immediate rescheduling of production. This should be no different in principle to responding to any other sudden change, although there will be a requirement that data, e.g. anticipated repair time, can be entered into the planning system by the maintenance people.

Scheduling manufacturing cells
Complex manufacturing cells, e.g. with multiple stations fed by multiple robots and therefore capable of creating internal queues, might require their own unique scheduling algorithms. These algorithms might include some of the classical OR solutions, such as the travelling salesman problem applied to an automated assembly station. However, some cell scheduling systems merely sequence the work given to them, without attempting to optimize throughput or to forecast completion times (see Chapter 10).

Referring to *Figure 7.1*, manufacturing cells would be linked to the CIM network and to the mainframe computers by connection to the shop floor ILAN. The question therefore arises whether the cell scheduling software would best be run on a mainframe computer, or on the cell controller; in either case it must interface with the factory capacity planning system, since that is responsible for forecasting completion dates for customer orders and needs to have good data about the progress of work in the manufacturing cell. This interface will be particularly important if work is given to the cell several batches at a time, with the scheduling system on the cell controller having a high degree of autonomy in the detailed planning. However, in some situations the mainframe systems may be able to make reasonable approximations about output from a manufacturing cell, and just leave the cell to carry on. Chapter 10 describes some of the extensive research work that is being done on this subject — another area within CIM where it is easy to hope that AI and expert systems will provide a solution in the medium term. Certainly, such machining and assembly cells are very expensive, and it must make good business sense to put a lot of effort into optimizing output.

A simple cell, for instance one machine tool and one robot, could probably be managed within the capacity planning systems described above, since throughput can be calculated on the basis of conventional set-up and run times (the robot cycle time being part of the set-up time).

However, in addition to capacity being available, the appropriate material and components must also be delivered. Within the context of CIM, there is a need to integrate capacity planning with MRP.

Integration with MRP

At a simple level the loop between MRP and CRP is as shown in *Figure 9.1*. This scheduling needs to be done to as accurate a timescale as possible, quite possibly to the nearest minute, since short-term variations in load may be important. Today's bottleneck might be tomorrow's idle capacity if work-in-process really is being kept to a minimum.

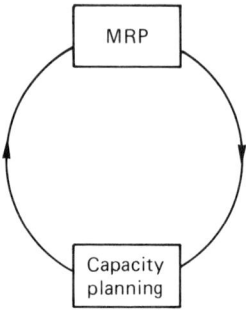

Figure 9.1 MRP and capacity planning in a closed loop

In principle, *Figure 9.1* can correct a major defect of capacity planning calculations, which is that important input data — the due date of each shop order — is probably wrong. If capacity is limited, then the shop order due dates calculated by MRP, using fixed lead times, are likely to be inaccurate because actual lead times will alter depending on queues of other work in the factory. Feedback of up-to-date lead times from the capacity planning calculations should help solve this problem. If customer due dates can be used directly in capacity planning, e.g. in a jobbing foundry, then the problem does not arise.

However, the situation is not as simple as *Figure 9.1* might imply. The data which MRP feeds down to the capacity planning system is are:

- quantity in each batch (from MRP batching rules),
- due date of the batch (from lead times off-set by MRP).

Routing data (see Chapter 5) must also be available. *Figure 9.1* indicates that the capacity planning system, after a finite planning run, will calculate expected completion dates for these batches. If these calculated completion dates (i.e. new lead times) are simply fed into MRP, and MRP then re-run on a net change basis, there may be little useful improvement to the schedule. If sufficient capacity is simply not available, then MRP (working from the Master Production Schedule) can only report that it needs components earlier than the dates calculated by the capacity planning programs, which in turn will report 'No can do' when asked to improve them, and the computer systems can cycle round *Figure 9.1* indefinitely.

Another factor that complicates *Figure 9.1* is that the feedback from capacity planning could take place at any or all of the three capacity planning phases discussed above (and in Chapter 5). A fully automated process factory would have most of the capacity planning done as part of the Master Production Schedule Planning functions; by the time the factory is in operation, there are few scheduling options left. At the other extreme of a jobbing machine shop, capacity planning is effectively done at phase three, when the work is actually committed by the customer. In between there is a wide variety of options; the most complex situations will require capacity planning at all three phases, as shown in *Figure 9.2*.

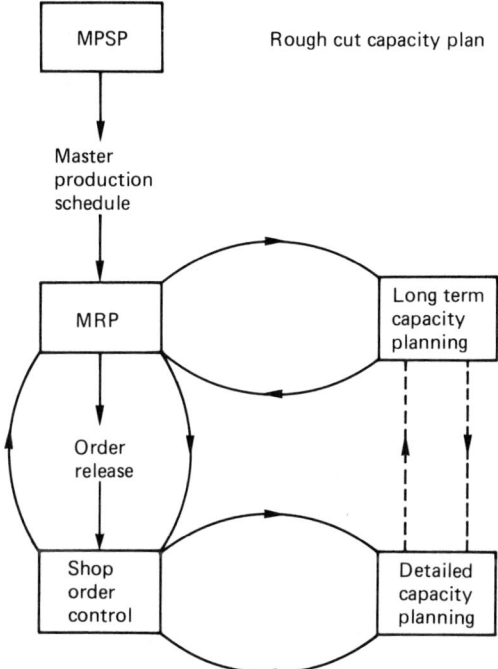

Figure 9.2 Multilevel integration of MRP and capacity planning

For both long and short-term capacity planning, some simulation of unknown future load will be required, since the effect of reducing work in process to a minimum and releasing jobs as late as possible is to reduce the actual load on the factory.

There are several means of improving the plan, which ideally could be invoked iteratively in the feedback cycles of *Figure 9.2*:

- arrange extra capacity, for example, overtime or subcontract,
- purchase some components complete rather than make them inhouse,
- split some of the batches (make part of the batch early to achieve the required due date, and the remainder later),
- use a 'cleverer' capacity planning package, which 'tries harder' to get the best out of the available capacity by use of better algorithms and more detailed simulation,
- delay delivery to the customer.

What combination of these options will be used in practice depends on the situation of each manufacturing company at the time. As already discussed in this chapter, it is by no means easy to identify where extra capacity is required, or which jobs should be subcontracted or bought complete to reduce demand on capacity. However, the prize which CIM offers, of the absolutely optimum production schedule, requires that the above options be automatically invoked, so that the feedback loops in *Figure 9.2* are stabilized; what defines an optimum schedule is discussed below.

The development from inaccurate capacity calculations to accurate integrated computer capacity planning may not be accepted easily (any more than were early attempts to keep stock figures accurately). It is nevertheless a probable consequence of the technical and market forces behind CIM. Considerable extra software function is needed to integrate MRP and capacity planning, and to run the factory automatically.

The factory autopilot

As discussed above, many current planning systems in manufacturing industry assume that human decision making will be available to sort out the 'details', and therefore that it is important for people on the shop floor to understand enough of the planning system's rationale to be able to second guess it. Furthermore, such systems are often oriented to getting the best out of the situation on the factory floor as it now is, rather than trying to plan the future. To exaggerate: 'Sales promise the dates the customer wants, Production makes the best of the ensuing mess'. In practice many plants are indeed run on targets determined with little analysis of how the production will achieve them, because the targets are fixed at MPSP level by considerations of market size and cash flow.

It follows that integrated, detailed material and capacity planning systems — the factory autopilot — might cause a real change in the management of production. Without detailed short-term plans being available, there has been every excuse to push the production scheduling problems down to the people working on the factory floor. CFM and JIT are easy to understand, hence they tend to get senior management support; even so there are examples of uncomprehending over-adherence to 'no work-in-progress buffers' causing visible reductions in output. The implementation of *Figure 9.2* will provide accurate capacity and material plans covering the next few months. Important decisions will be required on what to do about overloads and underloads. These decisions will be based on the accurate capacity planning systems described above, with batch sizes adjusted by expert systems, for example, and the results further modified by statistical extrapolation of yields and existing queue lengths. It will be less easy to leave the problems to be sorted, in the classical manner, when they actually hit the factory, and less easy to give the customer last minute excuses.

Moving the goal posts

It is always essential not to lose sight of the overall business objectives in discussions about the potential benefits of complex computer systems. Particularly, it must be clearly held in mind that at the leading edge of production technology many of the real world problems relate to actually making the product — perfecting the robot program, repairing the tool, maintaining a clean environment, etc. If the supplier is delivering the maximum number of components possible in the face of the yield problems on the production process, there is little benefit to be gained from a more precise capacity planning calculation. Also the excuse 'We know we can't do it because the computer says we can't', must be avoided. The human ability to find innovative solutions when under pressure must not be lost within the computer system.

Hence, however large the role that may ultimately be played by intelligent computers, the systems are never going to be perfect; they will continually need improvement. The need to involve people in analyzing business problems (discussed in Chapter 2) in no way lessens as CIM is actually applied, since humans will always take the major decisions. The fewer people there are in the factory, the more important that each of them adopt a positive, imaginative and coordinated approach to problems and opportunities. Techniques and training to 'make it happen' will be discussed below in Chapter 12. As has been emphasized in this book, CIM is a tool for producing better business results, not an excuse for unnecessary computer systems, however elegant and intellectually challenging they may be.

Hence 'moving the goal posts' might include simplifying the product, combining the Shipping Bay with Goods Receiving to save space, or subcontracting more work to limit the complexities of planning for limited capacity. The computer systems should not be programmed at great expense to agonize over how to split orders between suppliers if a better solution is to have just one supplier for that item. A better means of reducing work in process than thousands of lines of computer software may be to draw a small square on the factory floor, and thus limit the storage space to be used.

If, however, it is widely accepted that in most companies individuals cannot possibly schedule both suppliers and the shop floor to the precision that Kanban achieves in its special environment; the computer must therefore be available as an automatic pilot. It must be able to assess the plan produced by systems based on *Figure 9.2*, and then take decisions to improve it.

The measures of a good plan

The measures of success of the planning and scheduling system must be expressed numerically, or in some other way which is meaningful to the computer programs performing the decision analysis. Clearly these goals will at times conflict (for example cost minimization against most of the other goals). They will be defined from the company's strategic objectives discussed in Chapter 2 and will include:

- customer service, e.g. the percentage of orders being delivered to the agreed data,
- cost minimization. e.g. minimize overtime, subcontract work and airfreighting to make up lost time,
- consistency of information to suppliers and employees with few sudden changes to schedule,

- yield as a measure of quality,
- capacity utilization, to be high in expensive or constrained resources,
- actual product cost in relation to standard costs or contract price,
- various ratios of productive efficiency, e.g. actual process time divided by manufacturing cycle time to give a measure of speed of work through the factory, inventory turn ratio.

The development of computer systems that will accept such targets and automatically produce better plans is going to require some innovative work; academic work in this field is discussed in Chapter 10. However, manufacturing companies cannot wait. It is important to look at how to apply CIM architecture with present technology.

Applying the CIM architecture

Each manufacturing company will decide how it needs to manage the integration of MRP and capacity planning; effectively most companies are likely to need all three phases shown in *Figure 9.2*. The database and organization of the MRP programs was discussed in Chapters 4 and 5; the database could be distributed over different computers and be distinct from the programs that perform the calculations, with an interface as in *Figure 4.5*. The detailed work-to schedules will have records on the database such as are shown (simplified, and as if in a relational database) in *Figure 9.3*.

Job 1	Item ABC	Operation 10	Day 123	7 a.m.	w/c XY
Job 1	Item ABC	Operation 20	Day 123	11.30 a.m.	w/c YZ
Job 1	Item ABC	Operation 30	Day 124	1 a.m.	w/c XZ
Job 2	Item PQR	Operation 50	Day 123	11 a.m.	w/c XY

Figure 9.3 Simple database records for a shop floor plan

Whether much or little calculation has been required to achieve that schedule will depend on the circumstances. For instance, quite simple 'autopilot' software might identify that today's load is pretty light; it would just whip through the schedule and allocate orders in a few seconds (perhaps on due date priority) to produce the date and time fields in *Figure 9.3*. Tomorrow may be quite different (a large rush order, maintenance on bottleneck machine) and a large computer may spend hours trying different schedules, loading and unloading (in computer memory) to produce the plan. Whether linear programming, statistical forecasting, expert system, simulation or some learnt heuristic is applied could depend on the nature of the problem for that particular day.

This would be comparable to computer programs that play chess, in that the time taken to work out the next move is variable, depending on the situation on the board. The end result is always a simple chess move, however much or little analysis went into it.

The type of manufacturing (e.g. batch, continuous flow or process, as discussed in Chapter 3) will affect the planning and control functions of the shop floor systems. Batch production allows more flexibility and hence in principle need not be controlled in such detail as continuous flow or process production, although planning may be more difficult. Continuous flow and process lines are usually custom built, which implies that planning the throughput may be easier, since there is less freedom for scheduling decisions. However, quality and availability of components and processes must be controlled much more tightly than in batch production, because any hold-up may well stop the line and hence all production.

Execution of the plan

The output of the planning process (a schedule such as in *Figure 9.3*) will be made available to the shop-floor control systems. The robot or NC programs need to be prepared and tested. Even the most carefully checked plan may not be executable (because of machine or tool breakdown), so there should be an ability to interactively replan for these contingencies. Whether such replanning would be performed on a computer on the shop floor LAN, or on a mainframe, is a question of business expediency, not principle.

The data could flow to other networks. This communication might be via bridges or routers within one location; different locations could be connected via the mainframes or direct LAN to LAN connections via an X.25 link.

The operator on the shop floor will go to a VDU or workstation for the latest schedule. The windowing features and high function graphics of the terminal could be used to look at the CAD database for example, or check details of an important supplier order. The operator might have responsibility for producing robot and NC programs, in which case the program could be prepared and modelled on the terminal (see Chapter 8). Alternatively the operator might have more limited responsibility, and would look at a low-function ruggedized terminal for a display of schedule details, and then report back to the terminal using the shop floor technology described in Chapter 4. Such a terminal could be on the shop floor ILAN or on a loop which is attached to it. The logical next stage is therefore to look at likely relevant developments in the automation of manufacturing technology on the shop floor.

Further reading

BELT, B. (1987) 'MRP and Kanban — A Possible Synergy?', *Production and Inventory Management*, First Quarter, 71–80

BERRY, A. D. and DUHIG, T. (1987) *Integrated Project Control*, Maidenhead: Pergamon Infotech

HALL, R. W. (1983) *Zero Inventories*, Homewood, Ill.: Dow Jones-Irwin

PLENERT, G. and BEST, T. D. (1986) 'MRP, JIT and OPT: What's 'Best'?' *Production and Inventory Management*, Second Quarter, 22–28

SCHONBERGER, R. J. (1982) *Japanese Manufacturing Techniques — Nine Hidden Lessons in Simplicity*, New York: Macmillan

VOLLMAN, T. E. (1986) 'OPT as an Enhancement to MRP', *Production and Inventory Management*, Second Quarter, 38–46

Chapter 10

Today's automation and intelligent machines

Early examples of automation were typically dedicated to a particular task, i.e. the function was determined by the physical construction and set-up of the machine. This was only justified when the quantities involved in production were large. Difficulties of coping with product variation prevented the widespread introduction of automation to the manufacture of products in medium and small batch quantities. The challenges were two-fold: control and handling. Control involved instructing the machine to perform those operations specific to each different part or operation; handling involved locating and moving workpieces and tools in the workspace before, during and after processing. Understandably, progress was first made in industries such as aircraft production where the cost of individual parts and the need to machine increasingly complex forms provided sufficient justification.

Various attempts have been made to extend the application of automation by reducing the effective variation between parts. The 'family of parts' concept has provided the basis for a design and planning methodology called Group Technology (GT). Here similar parts, produced in small batch sizes are collected together into a group or family, all of which can be manufactured by a group of machines or 'cell'. As discussed in Chapter 3, these concepts radically affect factory layout. However, the move towards GT and cells has been slow and the reasons for this are discussed below. The application of cells has made the greatest progress in machining piece parts and certain types of machine tool have evolved as being particularly effective in the context of cells, principally the machining centre.

Progress in technology has led to the replacement of mechanical controls and cams, etc. by microelectronics. Information on the behaviour of the machine and the process is provided by means of sensors. This enables the process control loop to be closed, providing information on, for example, tool and workpiece position, process characteristics and machine condition, etc. Safety can be much improved with this technology. The control procedures are held in computer programs, of which several may be held in the computer's memory. The networking of controllers is discussed further below; various proprietary communications systems for factory environments have emerged, particularly from manufacturers of programmable logic controllers. Factory communications systems, of which the Manufacturing Automation Protocol (MAP) is an example, have been described in Chapter 7 as components of a CIM architecture.

The industrial robot has become for many the popular symbol of automation. The development of new robot architectures, new subsystems (particularly motors, sensors and controllers) and more powerful programming languages have led to robots being applied to an ever-increasing range of tasks. A more detailed discussion of the elements of automation in manufacturing follows, looking first at software aspects of programming and control and then at hardware.

Programming and control

NC programming

In the early days of numerical control, the programs were written in machine code; errors were difficult to detect and programs had to be tested before being used in production. Higher-level programming languages were developed (APT, CompactII, GNC) that allowed engineers to define programs in terms that were closer to 'English'. For examples of machine code and high level language programs, see *Figure 10.1*.

```
N081 G01   X–0045          F2500
N082 G04   H003                      S060M03
N083 G01   X–00051         F1000
N084                 Z–0125
N085       X00051
N086                 Z0125   F2500
N087       X–00052          F1000
N088                 Z–0125
N089       X00052
```

where N = sequence number, G = preparatory
    command, X,Z = coordinates
    F = feedrate, S = spindle speed,
    M = miscellaneous.

(a)

```
P4 = POINT/6.0, 4.5, 0.0  ⎫
LI = LINE/P1, P3          ⎪  Geometry description
L2 = LINE/P2, P4          ⎬
L3 = LINE/P1, P2          ⎭

        ⋮
COOLANT/ON                ⎫  Set-up commands
FEDRAT/6.0                ⎭
        ⋮
GO/TO, L1, TO, PL1, TO, L2  ⎫
GOFWD/L1, TO, C1            ⎪
GOLFT/C1, TO, L2           ⎬  Motion commands
GOFWD/L2, PAST, L3         ⎪
GOLFT/L3, PAST, L1         ⎭
```

(b)

Figure 10.1 (a) Fragment of NC code; (b) example of APT code

In addition to defining the relevant geometry of the part to be machined the NC program includes the sequence of machine operations. The programmer's job involves interpreting the drawings of the part and defining its geometry in a form that can be understood in the NC program. If the part has been designed using a CAD system, the geometry definition in NC programming can be bypassed provided that it is possible to represent the CAD geometry definition in a form suitable for CAM (Computer Aided Manufacturing). The transfer of part geometry from CAD to CAM is not a trivial task; the description of a part necessary to produce a drawing differs from that required to generate a program for machining the part.

The general availability of effective CAD/CAM, and the preparation of NC programs from a CAD system are comparatively recent. Even in 1984, some major turnkey systems failed to perform satisfactorily in this respect during an evaluation undertaken by a leading UK defence supplier. Now, at the end of the 1980s, the capability is well established and the programmers' task need only be to enter the machine operation instructions. Even this can be simplified by means of computer-aided process planning that defines the methods and tooling for the manufacturing processes available to the industrial engineer. While the engineer's job may seem to have diminished, the reduction in routine planning and programming enables him or her to increase productivity, become more constructively involved in the product design process (leading to 'design for manufacture' or 'design for assembly') or both.

Before a part program, written in a language such as APT or the output from a CAD system, can be loaded into a machine controller, it requires further processing to generate cutter location data from the geometry description, and then to translate it into a format appropriate to the machine and controller for which it is intended. The latter procedure is known as 'post-processing'. Post-processors can be expensive. They are required for every different machine and controller combination.

With some controllers, the operator can create NC programs at the machine. This technique is known as Manual Data Input (MDI). The operator uses the keyboard and graphics screen on the controller to describe the part and define the machining operations while the machine is still working on another part in 'background' mode. In more sophisticated systems, this is a 'conversational' process with the operator selecting options or answering questions presented to him by the controller. This approach can support multi-axis machines and is very suitable for toolrooms and subcontract machine shops where the overhead of CAM or CAD/CAM systems may not be acceptable. However there are limitations to the complexity of programs that it is desirable to enter in this way! As can be seen, the technology of the controllers has a significant impact upon how they are programmed and what tasks they can be expected to perform.

Robot programming

Robots offer a programmable, multi-axis handling capability that can be used to manipulate parts or tools. Programming robots can be a difficult task and would be impossible if the programmer had to define the movement and orientation of each joint throughout an operation. A solution has been to 'teach' the robot. This can be done by leading the robot to all the positions through which it should pass (while the motors are powered down). Each position is 'taught' to the robot, i.e. the current position as indicated by the position transducers (encoders or equivalent) is entered into the memory of the controller. A keypad or 'teach box' is used to enter speeds and other operational parameters in completing the program. Alternatively it may be possible to drive the robot from position to position from the 'teach box'. A demonstration robot 'teach mode' program for stacking is shown in *Figure 10.2*.

STEP	REM	FUNC	DATA	AXIS1	AXIS2	AXIS3	AXIS4
⋮							
205				38000	0	0	0
206				38000	11200	0	0
207		OUT	1				
208		CNT	3				
209				38000	000013	0	0
210		OUT	1				
211				55801	000001	985455	0
212	(FAST DOWN)			55801	109001	985455	0
213	(SLOW)	SPD	1	55943	10987	985490	0
214		OUT	2				
215				55800	10900	985451	0
216		OUT	2				
217		SPD	8				
218				55800	0	985452	0
219				55800	0	985452	0
220		CALL	681				
221				54200	0	984657	0
222				54200	10908	984657	0
223		SPD	1	54317	11000	984718	0
224		OUT	2				
⋮							

Figure 10.2 Extract from robot palletizing program

An alternative method of teaching programming involves an operator manipulating a light weight dummy, or 'teach' arm, through the required path. The machine controller records the path points at intervals and the program then drives the robot along the same path. This approach is particularly appropriate in paint spraying operations where it reproduces the skill of the manual operator.

'Teach mode' programming has several disadvantages, particularly that the program is not immediately intelligible to the operator, which can be a source of error. Debugging and program editing can also be tedious. Such systems often have very limited communications facilities for linking them with sensors or into larger systems. Typically they will only handle binary data though some have serial communications ports. Even so, not all systems are capable of handling conditional actions, i.e. able to select one of two alternative actions on the basis of an input value.

Increasingly robots are being installed in cells. A Delphi survey undertaken by the Society of Manufacturing Engineers indicates that by 1995 70% of all robots sold will be for use in integrated systems. Consequently controlling their operation has to be done in the context of other elements in the cell, machines, fixtures etc. To cater for the greater demands upon program capability, system suppliers have developed robot programming languages, such as VAL II, AML II and Karel, that allow structured programming techniques. As such they permit a greater degree of decision making within the program and are readily interfaced with sensors and other systems. A further advantage of these languages is that most of the programs can be written 'off-line' although typically the robot must still be taught critical positions. This minimizes the time for which the robot is taken out of production as the program can be prepared on a separate computer rather than on the robot controller. Typically the program will require to be tested by executing the program on a robot; however, there are systems available on which complete robot programs can be developed and tested in a model of the robot work environment (*Figure 10.3*).

These programming languages still require the programmer to break down the task into discrete robot instructions before preparing the program. Currently research is well established on the topic of task level

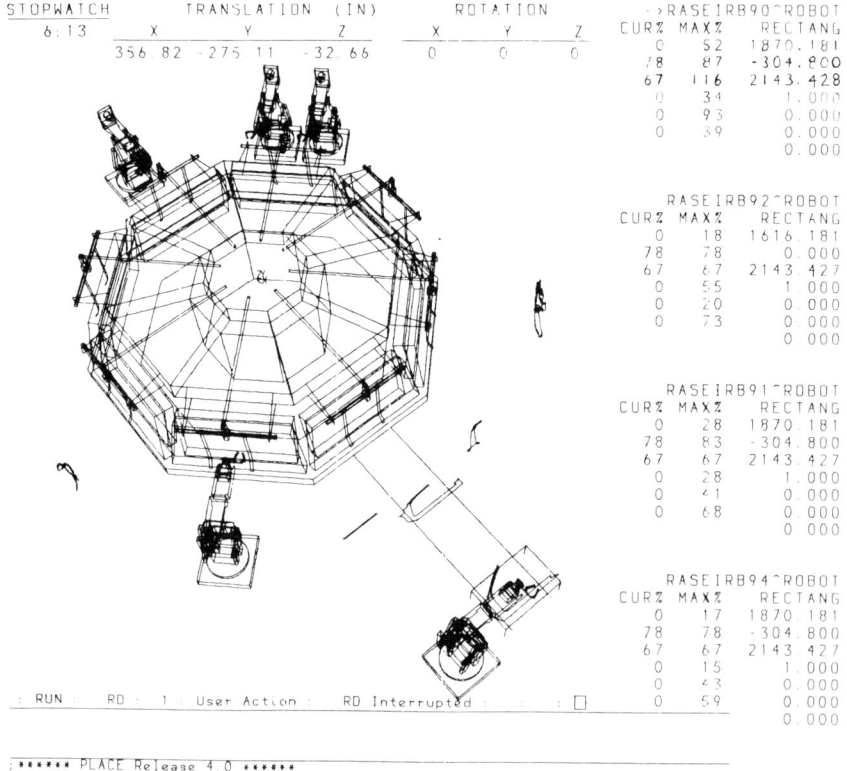

Figure 10.3 Computer model of a multi-robot cell. (Courtesy of McDonnell Douglas Corporation)

programming, where the engineer or programmer informs the programming system of the task to be performed, leaving the system to plan the operation and generate the necessary code to perform it. This approach will be discussed in the latter part of this chapter.

Cell programming

Cell programming involves programming a number of machines so that they will work together effectively. Depending upon the nature of the cell, it may be possible to use the controller of one machine to control the operation of the cell, with subordinate machinery functioning upon commands from the 'master' machine. High-level robot programming languages make it possible for robot controllers to be used effectively in this role. For some cells, however, such an approach is inappropriate and a separate controller is required. This might be a programmable logic controller, a minicomputer or a microcomputer, depending upon the application. The controllers will typically be programmed in relay ladder logic or conventional languages such as BASIC, Pascal, C or FORTRAN. With the growing emphasis on the cell as a building block of manufacturing systems, cell programming languages are beginning to emerge.

Planning

Planning the manufacturing method for any part or assembly is undertaken by a process planner or methods engineer. Some progress has been made in developing computer-aided process planning (CAPP) techniques. Systems generally belong to one of two types, variant or generative. Examples of the former include MiPlan and its derivative, MultiCapp (*Figure 10.4*). The characteristic feature of variant systems is that all parts are regarded as variants of one of a small number of master parts. Once the relevant master has been identified then deriving the manufacturing route and details of the method for any part is relatively straightforward.

Essential to variant process planning is an effective means of parts classification. The classification enables a part to be allocated to the appropriate master. An example is MultiClass, used in relation to MultiCapp; several classification systems exist, often deriving from Group Technology (GT) origins. Useful clarifications cannot always be achieved; an exercise in one plant machining prismatic parts for military radios showed that only 30% of parts could sensibly be allocated to masters. GT will be discussed in its own right below, especially with reference to the development of the philosophy of manufacturing cells.

Generative process planning differs in concept. Classification may be used to identify features on the part as before. The focus is on manufacturing rather than design, however. Instead of deriving the manufacturing method by reference to a master part, it is generated from first principles, extracting the appropriate operations and parameters from a manufacturing technology database. Some work has been done on techniques for recognizing the features of parts automatically; i.e. taking a CAD model and deriving the classification inside the program. Such systems are still the subject of research and interactive computer-aided

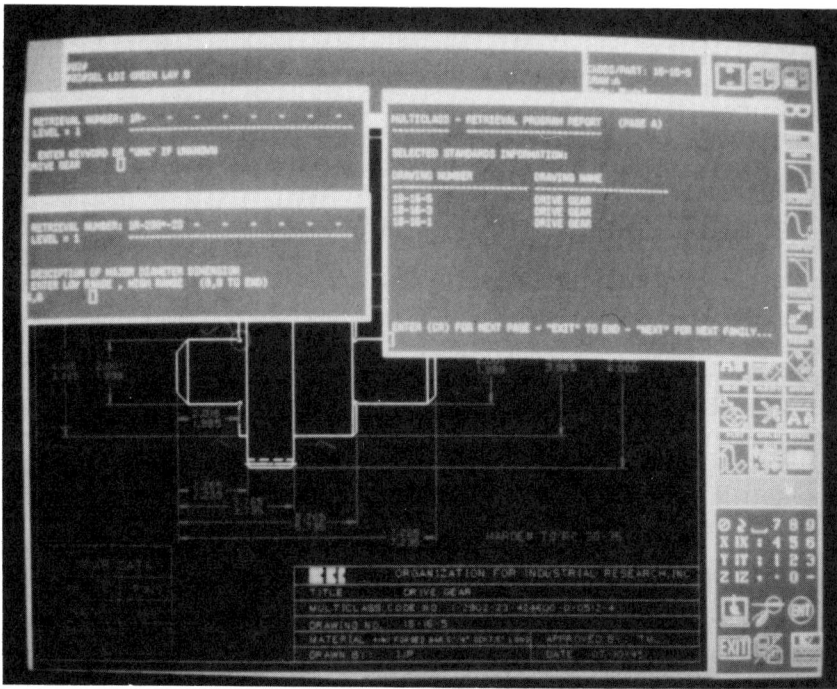

Figure 10.4 Operator's screen (Computervision Ltd)

process planning systems are a much more practicable proposition at present.

Group technology

The general importance of group technology now lies not so much in its practice as in the development of manufacturing thought which it illustrates. It represents a manufacturing method which identifies families of parts to be produced in cells of machine tools.

In principle, there are three approaches to establishing families of parts: visual inspection, production flow analysis, coding and classification. The most rigorous approach is by means of coding and classification that can reflect design attributes, manufacturing attributes or both.

Several alternative coding and classification systems have been developed. These are of two main types: polycode and monocode. In the former, also called chain code, the intepretation of each symbol is fixed. The code is readily intelligible but is inefficient, containing much redundant data. With mono- or hierarchical code, the meaning of each symbol depends on the value of the preceding symbol. It is efficient, but not easily understood by an operator. Given the relative strengths and weaknesses of the two systems, most practicable coding and classification systems are 'multiplexed', i.e. they incorporate both, employing monocode to hold permanent data (shape, function, etc.) and polycode for potentially variable data (batch size, annual requirement, tooling, etc.).

The benefits from group technology include: reduced tooling and set-up costs, reduced manufacturing lead time, reduced work in progress, standardization and families of parts. However the exercise of coding and classification is expensive; so is the physical reorganization of plant. Nor can these be dismissed as one-off costs since the possible requirement to revise the cell structure at some later date to reflect changes in the product mix must be taken into account. To monitor the alignment between cell structure and product mix, it may be necessary to establish rules regarding excursions from the parent cell. These are designers' rules which might, for instance, state that a design that required more than one excursion from the parent cell during manufacture would not be approved without the personal authority of the Chief Designer. If the rules become untenable that provides a trigger to restructure the cell.

Factory layout

Historically it has been possible to categorize factory layouts according to the nature of production, i.e. flowline for high volume and functional for medium size batch manufacture with machines arranged according to type: lathes in one position, mills in another, etc. One-off and small batch production is provided for by a group of functionally different machines offering a comprehensive capability such as may be found in a jobbing shop or a tool room. This is discussed in Chapter 3.

Functional layouts have disadvantages as they require: moving of parts between operations, balancing workload between sections, provision of buffer stocks. These lead to high 'floor times' (when parts are not being processed), high work-in-progress and long manufacturing lead times. Quality control is made more difficult. It is possible to optimise functional layouts using operational research tools and analyzing transport requirements in relation to the planned methods of manufacture. Computer tools to design layouts, and simulate their operation, facilitate this task. However, as mentioned above, physical reorganization is expensive and it is possible that a layout will be retained long after its relevance to the parts actually being produced has been lost.

Consideration of factory layout requires the understanding of hardware technology.

Machines for flexible automation

Automated management of machines within the context of CIM requires the ability to program the machine to perform a range of different tasks. The process of preparing programs has been described above. A machine and controller, with sensors monitoring the machine, tools and process make up the minimum functional configuration. Machines can have a process function, a handling function or both. Typically, cells will have both a process and a handling capability with machine tools providing the process function, industrial robots or automated guided vehicles the handling function. The distinctions are becoming blurred, however: robots can be fitted with tools that provide them with a process function while machining centres, turning centres and assembly machines often incorporate handling capabilities for parts, tools, etc. The essential features of machining centres, assembly machines, industrial robots and AGVs will be described briefly before considering how they can be integrated into systems.

Machining centres

Machining centres are required to be highly accurate, multiaxis, multifunction machines. They should be capable of milling, drilling and boring prismatic parts; the cutter spindle is vertically or, more often, horizontally oriented (in some designs it is possible to change from one orientation to the other). The machine justifies its cost by removing the need for many second operations (on a different machine) because it has the capability to produce accurately machined parts with a high surface quality that reduces the requirement for finishing operations, and multifunction capability that reduces the need for complementary machining operations.

They also need the minimum of operator supervision. For a discussion on machine design, see Weck (1984).

The machining centre's operation is controlled using a computer numerical control (CNC) unit. This incorporates drives for the spindle (powered by DC or AC motors) and the machine axes (usually DC). Closed loop control is employed almost exclusively with feedback from the driven axes being provided by linear scales located adjacent to the actuator, typically a ballscrew. This scale can be either glass, operating on Moiré fringe counting techniques, or of the electronic induction type. In some cases, encoders mounted on the motor shaft provide the feedback. Such systems can be described as semiclosed loop as they do not sense the actual movement of each axis directly.

Given the requirement for minimal operator intervention, machines must be designed for safe and reliable operation. Sensors should be built into each system to enable the controller to monitor plant, process and product. Monitoring the process includes tool length measurement, tool life monitoring, broken tool sensing, spindle load monitoring and automated work measurement. Increasingly, some level of adaptive control is being offered. Adaptive control for constraint is more effective in protecting machines and workpieces than in process optimization. The constraint approach ensures that machine operation remains within a predetermined range for critical parameters such as torque or power consumption, whereas optimization requires a greater depth of knowledge of the machining process and of its relation to tool and workpiece materials than is generally available.

Sensors fitted to the machine also monitor the correct operation of the lubrication and coolant systems, the pallet and tool changing mechanisms. Their output is fed to the controller for protection of the machine and workpiece.

The controller software requires diagnostic procedures to interpret the sensor data it receives and to decide on the appropriate response. The system should measure its response according to the severity of any error condition it discovers and to the system's capability for self-correction. For instance, excessive spindle power consumption may indicate a worn tool; the present operation should be completed at reduced speed and the tool replaced by a similar tool when next required. Indication that one axis had gone open loop, on the other hand, should result in immediate shut down of the machine. Note that the machine should have a safe shut down procedure.

In addition to CNC, the following equipment should be included with machining centres:

- A tool changer and magazine, typically with a capacity of 40 tools and sometimes of more than 100,
- a pallet changer or pallet loop to allow parts that have previously been set up on a fixture to be presented to the machine,
- a means of swarf removal, easily overlooked but not trivial,
- a coolant system, often instrumental in removing swarf and in minimizing temperature variations in the machine (for increased accuracy).

A system with this specification is capable of operating with the minimum of human involvement. The period of time for which it is selfsufficient will depend upon the size of tool magazine and how much work can be stored in the pallet pool.

The effort required to support the operation of a machining centre is significant. Apart from setting up fixtures (which in turn are mounted on

Figure 10.5 Flexible manufacturing cell. (Courtesy of KTM)

pallets), tool management is very important in keeping a machining centre working. Tool room support includes keeping the tools in good condition, setting them up for use, measuring them as set up and logging any offsets from the nominal position, supplying them to the machine as required and, if necessary, logging their life. In some companies, the complexity of the tool management system can match or exceed that of the machine and materials management. 'Design for manufacture' rules to limit the number of tools in use are usually required for systematic tool management to be practicable.

The indications are that machining centres are increasingly being installed as integral parts of flexible manufacturing systems or cells (*Figure 10.5*).

Industrial robots

Various definitions exist, of which perhaps that produced by the International Organization for Standardization (ISO) should be given precedence:

> 'The industrial robot is an automatic position-controlled reprogramm-able, multifunction manipulator having several degrees of freedom capable of handling materials, parts, tools or specialized devices through variable programmed motions for the performance of a variety of tasks. It often has the appearance of one or several arms ending in a wrist. Its control unit uses a memorizing device and sometimes it can use sensing and adaptation appliances that take account of environment and circumstances. These multipurpose machines are generally designed to carry out repetitive functions and can be adapted to other functions without permanent alteration of the equipment.'

Generally, robots can be categorized as one of five geometrical types. These are cartesian, cylindrical, polar, anthropomorphic (*Figure 10.6*) and automated guided vehicles.

Most robots are powered by electric or pneumatic drives, although hydraulic systems have been used in some heavy-duty machines. The particular capability of robots over other handling equipment is programmability. However, this feature is not always exploited; part variation is often low and batch size large. Reasons for this include the often high cost of tooling and feeders. In addition, early robot controllers often offered very basic facilities: point to point movement, sequential operation, a limited number of program steps and a primitive interface with sensors and other devices (capable only of exchanging data in binary format).

Advancements in microprocessor technology have provided powerful building blocks for controllers, particularly in the context of standard bus systems. These are described in more detail below. They enable complete control systems to be configured using various proprietary printed circuit boards, each supporting different functions (processor, memory, input and output, etc.).

The automotive industry has been a major user of robots and application areas in which they have been applied include spot and fusion welding,

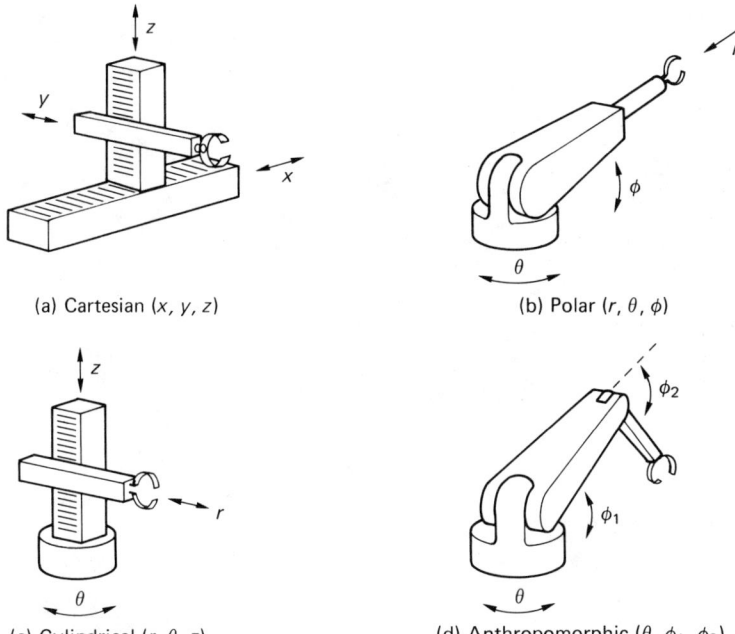

(a) Cartesian (x, y, z)

(b) Polar (r, θ, ϕ)

(c) Cylindrical (r, θ, z)

(d) Anthropomorphic (θ, ϕ_1, ϕ_2)

Figure 10.6 Alternative robot structures (SCARA robots are a variation of (d), mounted in the horizontal plane with θ fixed and sometimes having a z axis)

paint spraying (*Figure 10.7*) and various assembly tasks. Consistency of quality has been an important factor in justifying robots, as have been hostile operator environments and productivity. Indications are that the automotive market for spot and fusion welding systems may have become saturated and that future growth is likely to come in machine loading and assembly applications.

Assembly machines

The use of machines for assembly implies that either the volume is high, or that the assembly task is critical for instance in terms of cleanliness, accuracy, quality or the application of force.

Ideally assembly operations involve the placement of components onto a planar base: the placement of semiconductor devices onto a printed circuit board is a good example of such an operation. In printed circuit board assembly, the consistent placement of components at a high rate imposes stringent quality requirements on the design, on the parts to be assembled and on the assembly machine.

Mechanical assembly is typically a more complex, three-dimensional task especially if fasteners, springs or circlips are involved. Particularly in this context, close cooperation between the designer of the product and those planning its assembly is essential if effective automation is to be achieved.

All assembly machines require the feeding of the parts to be assembled and the removal of completed assemblies. Often the feeding of parts to the

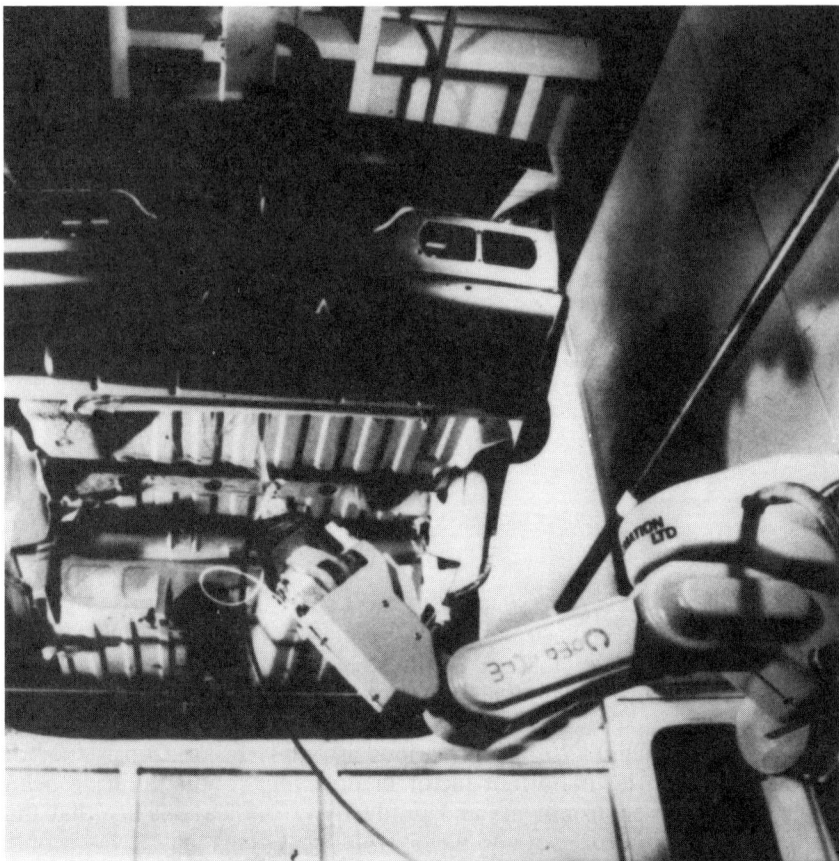

Figure 10.7 Automotive robot application: spraying underseal. (Courtesy of Rover Group)

assembly machine provides the most difficult problems in machine operation. Bulk supply requires separation, orientation and presentation of the parts, which may involve hoppers, magazines, bowl feeders or perhaps a conveyor. Parts may be difficult to separate; springs, circlips and washers are notorious in this regard. In such cases, reliability becomes a critical issue. As automated assembly becomes more commonplace, the 'condition of supply' in which components are delivered lends itself to machine assembly; examples are electrical components supplied in bandoliers and semiconductor devices packaged in tubes.

The nature of the placement mechanism will depend upon a number of factors: the arrangement for feeding parts, the elements of the assembly task, the physical envelope to be covered. For instance, if parts are supplied in an irregular pattern then a flexible handling system (i.e. a robot) linked to a sensor system for locating the incoming parts will be required. If handling is required only within a small physical envelope, then a simple dedicated mechanism may be the most cost effective solution.

An assembly machine will typically be modular in concept, with one or more input stages, a handling stage, an output stage and a controller. As far as possible, the modules or stages will comprise standard items, with perhaps some modifications for the application in hand. Both linear and rotational modules are available as catalogue items, ranging from modules for micropositioning with micron precision, to heavyweight handling equipment.

To be classified as flexible, the assembly machines should be capable of producing different products. This is practicable in the assembly of printed circuit boards where standardization in board and component sizes makes it possible to assemble products with different functions from parts with identical external shapes. Otherwise, assembly machines are often dedicated to a single product or product range (of which PCBs can be regarded as an instance). Greater flexibility is obtained from robot systems as described above.

Automatic guided vehicles

The first AGV system was installed in the US in 1954. The first European system followed some four years later in the UK. Early AGVs were essentially dumb units for transporting parts (and tools) in a manufacturing or assembly system. Primary directional control was effected by wire guidance. Commands regarding destination, permission to proceed, etc.

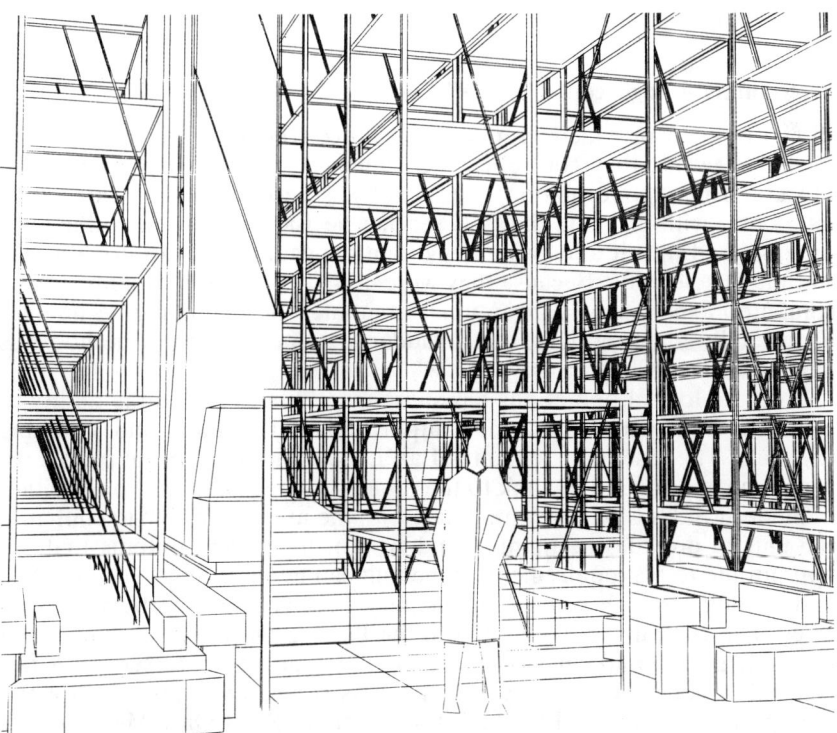

Figure 10.8 Automated Storage and Retrieval System. (Courtesy of T^2)

were transmitted to the vehicles from control boxes located at or close to critical junctions in the path. Extensive wiring was necessary, with the result that both installation and any modifications were expensive. Chain-driven tow carts that can be completely dumb have been preferred in many installations until recently. Other competitive transport mechanisms include the conveyor and stacker cranes.

Vehicle guidance continues to be provided by tracks laid on the floor. These tracks could be wires for electronic path sensing or a painted (or otherwise marked) line for optical sensing systems. Alternative navigation systems are being developed, however, for which no track laying will be necessary. In order to achieve greater design and operational flexibility and to reduce installation costs, as much of the control and communication function as possible is being designed into the vehicle. With increasing vehicle intelligence, AGVs are now a more practicable and attractive option.

In addition to transporting parts and tools between machines, AGVs have found application in Automated Storage and Retrieval Systems (ASRS) (*Figure 10.8*), and as mobile work stations, particularly for assembly.

Controllers

The controller types that are of particular relevance to manufacturing are discussed below.

Programmable logic controllers (PLCs)

PLCs originally provided a replacement for conventional relay controllers; in simple terms, they controlled the sequence of events in a machine or process. Binary information (i.e. on/off) is received from switches or interlocks; with this data the PLC reaches decisions according to the logic programmed into it and generates outputs (again binary) to drive motor starters or solenoids. Because they were used by operators with little or no computer experience, conventional computer programming techniques were totally inappropriate, and a ladder-logic programming method was developed (see Chapter 4).

To these logical operations has subsequently been added the ability to calculate, analogue input and output capability, timer and counting channels. The essential elements of a PLC unit are the central processor, program memory and the link to the process or machine. The construction of modern PLCs is typically modular, with the user being able to configure the unit to provide the number and type of inputs and outputs that are appropriate to his requirements. One unit can monitor hundreds of inputs and outputs, though some thought must be given to the effect on scan rates. The central processing unit in the PLC can scan a given number of channels per second; in simple terms it can be set up to service a few channels frequently or many channels less often. Program memory can now be in the form of PROM (Programmable Read-Only Memory, see Chapter 4) modules which are replaceable, allowing the user to load one of several programs if the machine or process is multi-purpose. Optional modules include printer interfaces which enable the PLC to produce reports on the machine or process it is controlling.

Proprietory systems for linking PLCs together in a factory are offered by many PLC vendors. Provision has also been made to link PLCs to computers; the personal computer (or, confusingly, PC) is frequently being used for this purpose because of the wealth of software available for it at comparatively low cost. The PC can for instance analyze data from the PLC to provide production or quality control.

In parallel with these developments, programming methods have improved to facilitate the full exploitation of PLC capabilities. This reflects the increasing knowledge of computer technology in the factory; versions of BASIC are now available to program PLCs which, as indicated above, can themselves offer many of the facilities of a microcomputer based system. Consequently, PLCs have been finding increasingly wide application in industry for machine control, drive motion control for DC servo or stepper motors, material handling, test station control and paint colour control. A further application is in cell control where the PLC supervises the operation of a number of machines, each of which is itself computer controlled.

Numerical control

Early controllers read programs on paper or magnetic tape, one block at a time, intepreted that data and drove the machine tool accordingly. Tape readers were critical elements in the system since the data was transferred from them via a single block buffer to machine control. The consequences of error were immediately fed through to the machining process with the risk of scrapping the part being machined. Correct programs were no guarantee of correct operation since tape and reader failure could introduce errors at any time. Because NC could only be justified where machining costs were high, the cost of scrap or at best salvage was often high also. The tape and tape readers fell short of the high reliability requirements of this application. The first approach to removing them from the system was the introduction of direct numerical control or DNC.

DNC comprised one or more NC machine tools linked to a host computer, typically in a star network. The program was downloaded to the NC via the computer link block by block bypassing the tape reader. As far as the NC unit was concerned, it was operating normally. Units where the remote host is functionally equivalent to the tape reader are known as Behind the Tape Reader or BTR systems. This overcame the reliability problems associated with the tape and readers. A subsidiary benefit was that tape handling, which could be time consuming, especially with long programs, was no longer necessary. This approach had two significant disadvantages: cost and vulnerability to computer failure. Because the host had to feed programs to each machine block by block on a continuous basis, there was a heavy communications demand on the computer which limited the number of machines it could service. Typically a DNC system would include four or perhaps six machines. Each host computer therefore represented a very significant cost overhead on the machines in the system.

System reliability depended upon the host or communications network; failure of either of these elements resulted in machine stoppage and possible scrap. Few DNC systems were implemented and alternative approaches to improving system performance have emerged.

Computer numerical control

In computer numerical control (CNC), the hard-wired logic in NC units is replaced by a programmable computer. The controller is therefore more flexible, with the result that interfacing it to a machine is correspondingly easier and cheaper. The first advantage of a CNC unit is that it can read NC programs from tape and store them in its own memory. Provided that the memory can hold a complete program (increasingly probable with the decreasing cost of memory) then the tape need be read only once, the program can be tested cutting air or low cost material and then machining of the part proper can proceed. The reliability of memory is such that failure is extremely improbable. With the reduction in cost and size of computers, and the rapid increase in their power, CNC equipment has become progressively cheaper and more powerful. By the later 1980s, systems at 20% of the price of those available 20 years earlier offered incomparably greater functionality.

The computer within the CNC system has led to the provision of computer-like peripherals externally. These include a CRT screen, keyboard and in some instances, floppy disk drives. For the operator, this has provided the opportunity to gain extensive information from the controller regarding the machine and process. The operator can select one of several screen displays that will give him information on the current job: for example, a display continually updating the coordinate positions of each axis or the position in the program. Other programs in memory can be reviewed or a program input at the controller, which at the same time is running the machine from another program. CNC units also offer a self-diagnosis capability. With this, the controller can report on the screen problems with any of the machine and control functions such as tool changers, servos, spindles, coolant systems, lubrication, etc.

With the availability of CNC equipment, a second generation of DNC systems has emerged. The concept of distributed numerical control provides for a host computer serving several machine tools as before. However, if the machine tools are equipped with CNC units, the host can transmit complete programs to each CNC, which can then operate autonomously with its machine tool until the program requires to be changed. Given the intermittent communication load between the host and any given machine, a single host can serve a significantly larger network of machines than would have been possible with DNC. Moreover, host or communications failures do not necessarily cause the system to stop since each machine is capable of operating in 'stand alone' mode.

Factory communications requirements extend beyond NC to production and general management. DNC therefore should therefore no longer be considered in isolation but in the context of an overall communications architecture (as discussed in Chapter 7).

Industrial computers

In addition to PLCs and CNCs, industrial computers are establishing a direct presence on the factory floor as elements in inspection systems and for process control. Typically, the computer will be linked directly or indirectly to sensors that are measuring some aspect of the product or of process behaviour. A wide range of computers is available, ranging from single-board computers with a limited interfacing capability but allowing compact installation to bus-based systems offering powerful processing potential with extensive input and output capability.

Many vendors offer systems based on bus specifications. Bus interconnection systems provide a basis for building up computer systems from a selection of standard boards. In this way, the specification required for particular applications in terms of processor, memory, input and output capability, communications and peripherals can be achieved without designing and making special purpose boards. Increasingly, builders of manufacturing systems are using these modules as the basis for controllers and for analyzing sensor data. A brief overview of the types of boards available on these buses is given in *Figure 10.9*. Consideration will now be given to the sensors that provide the interface between all control systems and the process or machine under control.

Sensors

Sensors detect, count, measure or identify. At the simplest level they may confirm the presence or absence of an object of interest; for instance, a component to be assembled or a tool for a machine tool. Alternatively a sensor may count the output of a machine, being triggered every time an item passes it. Other sensors provide measurements of the variable they are monitoring, for example a thermocouple. The function of a sensor may be affected by what is done with its output. This will be discussed below as various different types of sensor are considered in more detail.

- *Microswitches:* mechanical devices in which the movement of a plunger makes or breaks an electrical contact. The switch has open and closed positions and an allowance for overtravel. Several different mechanisms are available to actuate the plunger, enabling microswitches to be configured for a range of applications, for instance providing unlimited overtravel.
- *Proximity switches:* non-contact devices relying on electrical and magnetic properties to change device behaviour. Typical of these are inductive and capacitative devices; the former are more common for industrial applications and typical sensing ranges are from 1 mm to 150 mm (note that as sensing range increases, so does sensor size while the maximum switching speed drops). This approach is not suitable for detecting non-conductive materials.
- *'Opto' switches:* non-contact devices incorporating light sources and sensors and with which the interruption of the beam between source and sensor indicates the presence of an object. By adjusting the beam diameter and sensor area, resolution of 0.025 mm can be achieved.

238

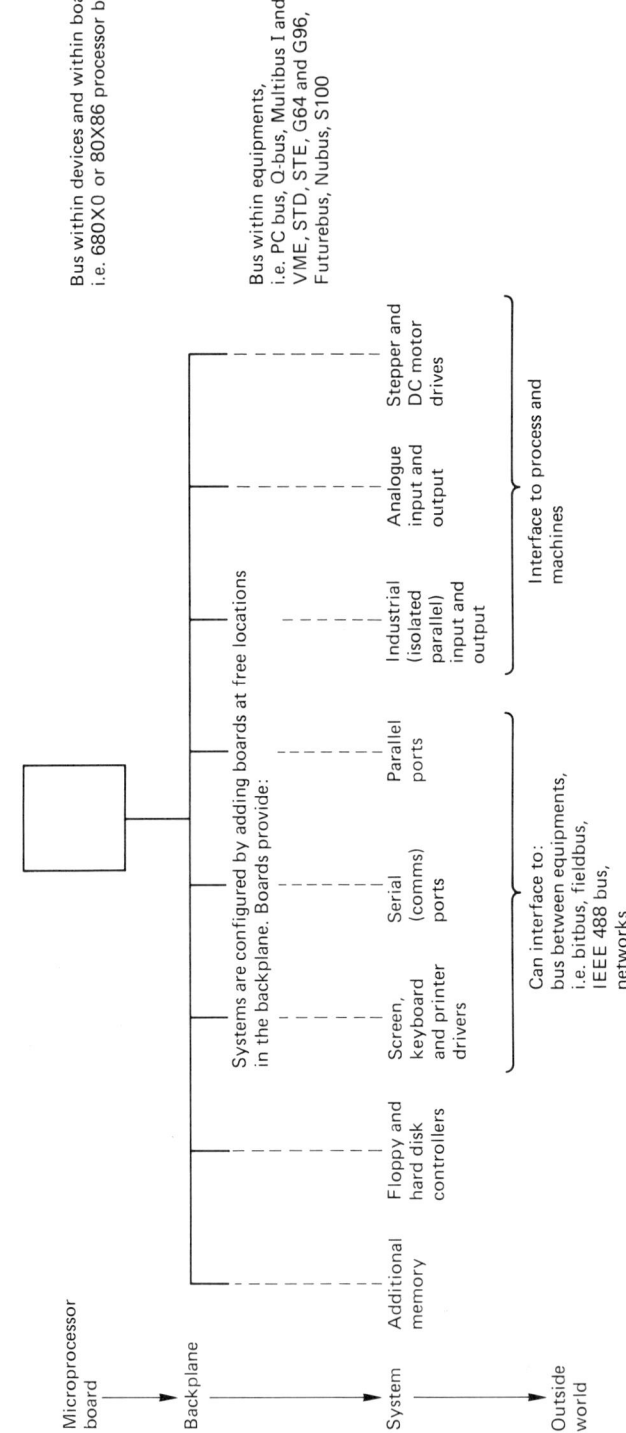

Figure 10.9 Industrial micro systems

'Opto' sensors are versatile and they find wide application as detectors in automated systems.

General-purpose sensors such as these can be found in machine tools, assembly machines, robot systems, packing lines. Wherever the satisfactory completion of an operation depends upon the presence of an item, be it a tool, a component or even a label a potential 'detector' application exists.

One sensor that has found particular application in automated manufacturing is the touch trigger probe. It is a high precision device that is triggered by contact from any direction; a deflection as small as one micron can be sensed. Software for operating the sensor is available from many controller manufacturers and it is used in machine tools for setting up work (less appropriate in automated systems), part program offsetting, tool checking, tool wear measurement and for part inspection. Measurement is achieved by using the sensor in conjunction with whatever transducers indicate the position of the machine axes; accuracy is largely determined by the machine's specification. The sensor can be mounted on the machine bed or in the spindle, as is most suitable for the intended use. In addition to machine tools, touch trigger probes are fitted to inspection systems (their original application), both in the form of coordinate measuring machines and robotic systems.

Sensors that count are simple detectors, provided with output signal conditioning systems that can sum the number of times the detectors are activated. Many linear and rotational transducers adopt this principle, being based on the counting of optical, magnetic or electrical pulses.Incremental and absolute encoders, synchros and resolvers fall into this category as do Moiré fringe-based systems. Encoders or linear scales typically provide the feedback for controlling machine tools and automated machines. Tachogenerators (measuring velocity), potentiometric devices and displacement transducers output a voltage dependent upon the measured value.

Other sensors give a direct measurement. These include force transducers, which are often specified in automated machining systems to detect actual or impending tool breakage and to measure tool wear. Certain sensor systems for lathes employ strain gauges built into the machine spindle bearings and provide a dynamometer for measurement of cutting forces in one or two axes. Production machines tend to employ spindle power measurement as a means of monitoring the cutting process and tools. One implementation of this approach collects data from the spindle drive (current, voltage and rotational speed) and calculates the drive power and torque. Corrected for 'windage', the torque is compared with six predetermined limits permitting some measure of adaptive control, detecting surfaces, turning on coolant (i.e. detects when cutting starts), maximum power overload, cutting tool protection and machine tool protection.

A further method of process monitoring is Acoustic Emission (AE). AE results from deformation in the cutting zone; signals generated by a chipped or broken tool can be distinguished easily from what is output in normal cutting conditions.

Optical techniques are widely employed, ranging from the simple optical sensors described already to machine vision systems that can detect, count, measure or identify.

Machine vision

Machine vision systems comprise a camera, usually solid state, some interfacing hardware and a computer. The sensors in solid-state cameras are, in effect, a two-dimensional array of photosensitive elements (the separations can be as small as 13 microns). Most commonly, they are arranged to support the horizontal and vertical line pattern of domestic TV since commercial demand and economies of scale result in this being the most cost-effective solution. On exposure to light, each element will become charged in proportion to the intensity of the light incident upon it. The charge is an analogue representation of the light input and must be converted into digital form before being transferred to memory in the computer system. A large memory capacity is required to hold the data from each frame (or picture), its size being dependent upon the number of elements in the array and how finely the light intensity must be resolved, typically to one in 256 parts. This part of the system is often called the frame grabber.

The next stage in the process is to reduce the quantity of data to manageable proportions. This is achieved by extracting from the frame the essential features. Typically, this may involve identifying boundaries of objects through thresholding or edge detection. Thresholding reduces the image to dark and light areas by classifying all elements registering a certain charge or more as light, all others as dark. Edge detection is a form of image enhancement, techniques for which range from the relatively simple to the very sophisticated. Image enhancement algorithms can be implemented in software but they represent a very considerable data processing load. If real time image analysis is required, for instance in process checking of printed circuit board assembly, then a software based system would require a dedicated computer. Many of the basic data processing operations have been implemented in hardware, thus minimizing the load on the central processor and enabling practical systems to be based on 16 and 32 bit microprocessors.

Once an acceptable image has been obtained it can be compared with master data to identify parts or to inspect them. Object recognition may be based on the area of an object, its perimeter, the position of its centre of gravity, etc. Alternatively, pattern matching techniques may be used to inspect known objects, for instance for the presence of all components in printed circuit board assemblies or of features in machined parts. Regular location of the parts to be identified or inspected simplifies the image analysis considerably. The provision of consistent lighting is often an important factor in the engineering of machine vision systems.

Not all sensors in vision systems are two dimensional; many applications are better served by a linear array of light-sensitive elements. The basic technology is the same as that for area arrays but the arrangement of elements is one dimensional. Sensors with up to 4096 elements in a single array were available by the late 1980s; again, availability has been improved as the result of a commercial requirement for the sensors, this time as components in facsimile machines and digitizers. Industrial applications of linear arrays include measuring elements in inspection systems, measurement and flaw detection in web production processes such as steel, paper, plastic film, etc. Processing data from a linear array is simpler than with area arrays as fewer elements are involved. As before, the data can be handled in binary form (thresholded) or in grey scale (a proportionate value) — most industrial systems are monochrome.

Vision is one of the areas where the techniques under development in research laboratories will find increasing application.

Intelligent machines

The growing discipline of machine intelligence will have a major impact on manufacturing automation, particularly developments that are applicable to discrete parts manufacturing.

Machine intelligence is intended to allow the constraints on the systems of the future to be relaxed, to have the ability to cope with uncertainty and to require less dedicated engineering. This technology is necessarily speculative. The approach includes the creation of intelligent systems with minimal programming, since such programmable systems have a high software complexity (and cost) as is indicated by *Table 10.1*.

Table 10.1 The software overhead in complex robot-based automation systems (after Carlisle, 1985)

University of Michigan generic part sorting cell	30 000 lines 4 cpus
Adept demonstration at Robots 8 exhibition	6000 lines 2 robots 2 vision systems 1 conveyor
Martin Marietta intelligent task automation project	75 000 lines 8–10 cpus 5–6 languages
Schaffner electronic assembly system	18 000 lines in robot controller and host PLC 10 other PLCs

Machine intelligence, artificial intelligence and intelligence

Machine intelligence has grown out of the older discipline of Artificial Intelligence (AI – see Chapter 7) and work in 'advanced robotics'.

AI simulates cognitive processes using computers. In the early days, AI attempted to understand human intelligence ('the relating activity of mind', 'insight', 'the capacity to meet novel situations with new adaptive responses'). Increasingly problems such as the building of autonomous systems are being treated as complex information processing tasks.

'Advanced robotics' is essentially reasoning about sensor data and using the results of this reasoning to perform a useful task, especially in an unstructured environment. This can be seen in research areas such as task planning and image understanding. Task planning, for example, generates the necessary robot paths to assemble two objects from a high level abstract task description, such as 'assemble A and B'.

Machine intelligence in the manufacturing domain extends this, in usually a more structured environment, to systems that are capable of accepting instructions from an external 'automatic' system, such as a management or CAD system. To achieve this in practice requires two extra technologies to reach the shop floor: cheap and powerful computing plus inexpensive and rugged sensing technology.

Production rule systems

An expert (or knowledge based) system is software that uses knowledge and reasoning to perform complex tasks usually associated with an expert. These are the most encountered element of AI based technology.

Expert systems are generally production rule systems. A production rule system consists of a rule base, a short term or working memory to hold the current system state and a rule interpreter. This interpreter (the inference engine) selects one rule (having resolved any conflicts between rules) and applies (or 'fires') it.

As described in Chapter 7, the rule base of such a system is a set of IF (antecedent) THEN (consequent) rules. These rules, the method of knowledge representation, are often written in a 'declarative' language. Declarative implies a more English like, 'what is' representation of the information rather than the procedural 'how to' representation associated with the conventional computer languages like FORTRAN and BASIC. *Table 10.2* shows the structure of a production written in PROLOG together with English-like translations (see Chapter 7).

Table 10.2 A production in PROLOG

mother(X,Y) :–
   parent(X,Y),
   female(X,Y)

[For all X and Y,
   X is the mother of Y (consequent) if
   X is the parent of Y (antecedent) and
   X is female (antecedent)]

[IF X is a parent of Y and female (antecedent)
   THEN X is the mother of Y (consequent)]

The IF–THEN, antecedent–consequent, condition–action, state transition construction of such rules is becoming widely applied in many of the areas discussed below which will cover current machine intelligence issues and achievements. Two very forward looking areas, the intelligent manufacturing system and the intelligent manufacturing workstation, will include some approaches to planning and scheduling. Advanced robotics work demonstrates parts of such systems, such as the autonomous vehicle for manufacturing, task-level robot programming for assembly and image understanding.

Intelligent manufacturing systems and distributed manufacturing systems control

The next major innovation in manufacturing systems could well be intelligent manufacturing systems (IMS). The intelligence in such fifth-generation, flexible factories is the ability to adapt quickly and economically to unpredictable changes in requirements. Such systems will be capable of solving unforseen problems, even with incomplete and imprecise information.

This work has been made possible by the technologies of decentralized/ distributed multi-microprocessor control systems (linked by fast local area networks) and programmable freely routable vehicles and the software technologies of AI.

As has been discussed, most conventional automated factory control systems are devolved hierarchically to keep the control problem manageable. A particular model for this to generate intelligent behaviour was developed at the Advanced Manufacturing Research Facility (AMRF) at the National Bureau of Standards in the USA. In this model, high-level goals are successively decomposed by lower-level control modules until a sequence of coordinated primitive actions is generated. These primitive actions will be simple machine-level commands which can be understood by, for example, a proprietary robot controller or other shop floor controller. Each control module is implemented as a state table, each line of a table effectively representing a condition–action rule.

Much of the system level work in intelligent manufacturing explores the possibility of controlling manufacturing systems without employing such strict hierarchical control strategies. The thrust of the argument for this is that hierarchical control could be replaced at all levels of a company's operations by autonomous cooperating sybsystems that work together to achieve a corporate goal.

This philosophy is based on a 'market-place' model where autonomous intelligent entities interact to satisfy their needs. There would be no supervisor in such a system; all of the entities would be equal in the negotiation process to obtain services and customers, and would cooperate with each other to obtain mutual satisfaction.

Cooperation as a control strategy requires that devices can operate independently of the rest of the system and are able to refuse requests for service from other devices, whilst still being integrated into an effective system.

There has been considerable research on different approaches to cooperation at the machine level in manufacturing systems. In order to introduce a 'market-place' model at this level, the parts must be intelligent also and have their own dynamic information storage and processing unit.

A comparison needs to be made between hierarchical and nonhierarchical cell control. This involves modularizing system control and reducing global information by locating decision making where the data originates. Heterarchic 'equal' systems can have smaller control programs that are easier to understand and only refer to local data (compared to the single large database approach). To determine the relative performance of different control schemes a team at the University of Wisconsin–Madison, for example, have constructed an experimental cell and tested three different strategies: centralized control using a single processor to sequence the cell operations and interact with the device controllers; hierarchical control using a supervisory controller and three slave processors to handle communications with devices; and heterarchical control.

For heterarchical control, each station and part was programmed as an intelligent entity that used the communication network to arrange transactions with other entities. A part would broadcast a request for a certain type of machine repeatedly on the network until it received a response from a free machine. It would then send a reservation message to the machine, which would confirm the reservation by sending an acknowledgement message. The part would then arrange with the robot to be transported to the reserved machine, and transmit its part data record to the reserved machine. A part program, running on any of the processors in the cell, would thus negotiate its way through its required manufacturing steps by cooperating with the robot, CNC machines and other stations.

The control software effort required for the heterarchical system was reduced significantly compared to the other two systems when the number of lines of control code are taken as the basis for comparison (*Table 10.3*). Extension of the system is also simple as it is only necessary to add another entity program to the network rather than edit a central program. The system was also seen to be inherently fault tolerant as faulty machines simply took no part in the negotiation process.

Table 10.3 The Wisconsin–Madison results

	Centralized controller	Hierarchical controller	Heterarchical controller
Lines of source code	680	2450	259
Software development cost ($)	17 000	61 250	6475
Expansion software cost	17 000	960	0
Machine utilization	—	64%	60%
Memory requirements	—	50 608	7104
Average CPU utilization	—	20%	60%
Complexity	low	highest	lowest
Flexibility	lowest	high	highest
Modifiability	lowest	high	highest
Fault tolerance	lowest	low	highest
Intelligent parts	lowest	no	yes

A team at Purdue is doing similar work with a different method of implementation. The work is targeted at the fast turn around of orders for replacement parts for machining equipment in batch sizes sufficient only to fill the order, i.e. there is no production for stock and batches are consequently very small. This system involves intelligent parts and intelligent workstations that perform the manufacturing processes. Parts hold their own quality control history, performance measures and due dates. Workstations hold NC part programs and information on processing capability, production capacity, maintenance records and tooling management.

The system is 'object-oriented' in the sense that the parts and workstations can be viewed as objects that pass information via messages and keep private some of their internal methods and data. Parts communicate with workstations to determine current load, processing ability, historical quality characteristics, and current estimated completion time for the required operation. Workstations communicate with parts to determine type of operation to be performed, type of material, and part's priority. Each object must have its own advanced computing capacity but this need not be physically located with the object.

A common memory 'Bulletin Board' is used to manage the procedure. Parts post required operations in the part of the bulletin board set aside for that type of operation, and the bulletin board notifies the workstations that can perform these operations. This decouples the parts from the workstations making the system robust to changes in available machines, which are added or deleted from the system simply by notifying the bulletin board.

The intelligent manufacturing workstation

Much of the work discussed above presupposes the development of an intelligent workstation (the manufacturing workstation carries out production processing; this must not be confused with the computing workstation which carries out information processing) or cell that is able to execute the planning, machine coordination and monitoring tasks currently carried out by the engineer and manufacturing craftsman. It is envisaged that all these tasks will be carried out at the machine itself from a product description.

The basis of much work on this subject is to divide the task carried out in the intelligent workstation into time domains. The longest of these, with time intervals measured in minutes, is that of the engineer. In this domain the primary interest is planning; process planning and path planning for cutting tools and manipulators, for example. The intermediate domain is that of the foreman, with a time interval of seconds. Here the task is to coordinate the actions of many automated devices. The fastest domain is that of the manufacturing craftsman who is carrying out the realtime task of process monitoring.

Planning in the manufacturing sense covers such aspects as process selection and ordering, robot and NC path planning, fixturing and grasp, how objects are gripped, planning. Knowledge-based planning systems share many techniques and problems in common with expert systems. The production rule-based expert system generally works in a convergent fashion to narrow down a wide range of initial possibilities. Planning on the other hand is a 'synthetic' process of piecing together subtask descriptions into an overall plan. Such a synthesis requires inherently more search, which is complex and computationally expensive even when reduced by knowledge-based heuristics, because there are alternatives to consider and conflicts to deal with. The operation of a hierarchical nonlinear AI planner proceeds as follows. Based on a hierarchical representation of the plan, a set of goals or a skeleton plan can be given and expanded out to greater detail. The planner searches through alternative methods of expanding

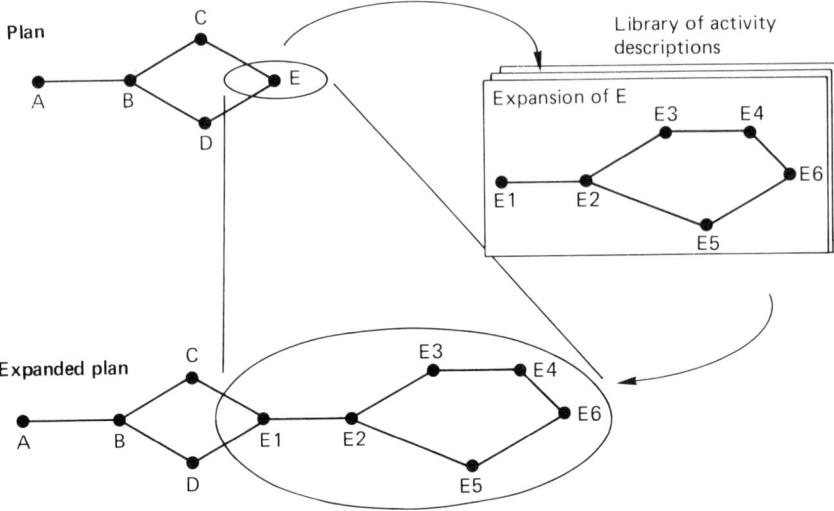

Figure 10.10 Expanding the step of a hierarchical plan

plans to lower level ones (as shown in *Figure 10.10*) and corrects interactions between solutions to different parts of the plans. At each level the plan is represented as a network of nodes in a form that allows the use of knowledge about the problem to restrict the search for a solution.

An element of the planning domain for the workstation can be indicated by work at Carnegie Mellon University. This is a program called Machinist that plans the set up of raw materials for machining. Machinist is part of a larger project, The Machinist Expert, the long-term objective of which is the total automation of manufacturing processes for small batch production. This particular element will produce a plan of an ordered sequence of global set-ups. Each set-up will specify an orientation for the part, a fixture, and an ordered list of 'features' to be cut into that part during the set-up. Each feature has an operation and tool associated with

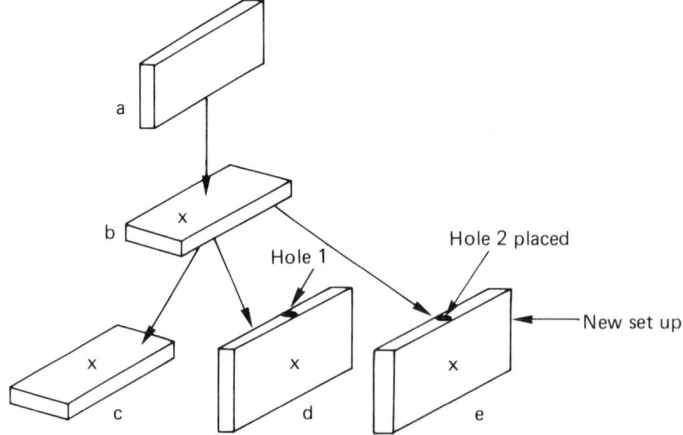

Figure 10.11 A modified global plan for a machining operation

it. A feature is an individual geometric shape to be cut into a part. The implementation produced so far can handle parts produced on a machining centre from rectangular stock with 2.5-D features and is particularly directed at very small batch manufacture. Examples of plans produced by the system from an original plan template are shown as *Figure 10.11.*

These approaches generate a completely worked out plan, which is then executed. Another interesting approach to a similar planning and execution problem is that of opportunistic scheduling, which exploits executional uncertainty. In any system there will be unforeseen or fortuitous events that occur, and these can be used to advantage. A plan with only partial ordering (events that necessarily follow each other) which allows a number of options at execution time is pruned by data that arrives during the time the schedule is being executed. This work has been tested by applying it to the building of a gearbox (*Table 10.4*) to show that this approach is more successful than other scheduling approaches in situations of organizational uncertainty, where for example different parts of the gearbox arrive at different times.

Table 10.4 Experimental results of strategy comparisons for the assembly of a gearbox (units: average robot operations used)

	Fixed strategy	*Opportunistic strategy*			
Parts visible	*Buffer single sequence*	*Build single sequence*	*Buffer single sequence*	*Build partial order*	*Buffer partial order*
1	21	57.1	18.0	29.5	16.3
1 or 2	21	37.8	17.0	23.1	15.0
1 to 5	21	20.8	15.4	15.1	13.65
1 to 10	21	17.25	14.5	13.8	13.05

A Cell Management Language (CML), which provides powerful generic software tools for the rapid integration and coordination of multivendor equipment into working cells, has also been developed at Carnegie Mellon. CML is not primarily concerned with the standardization of message form and the transfer of messages across particular transport media, but with the understanding and generation of messages produced by existing equipment from many manufacturers.

In order to understand all of the different manufacturers' languages, CML includes the facility to build an interpreter, or 'machine-specialist' for each device in the cell. These are built by defining the grammar of each of the individual machine languages in table form (see e.g. *Table 10.5*). This

Table 10.5 A grammar for simple declarative and interrogative questions

Grammar	*Field 1*	*Field 2*	*Field 3*	*Field 4*
Sentence	Grammar:Question	Grammar:Declare		
Question	verb/label="−quest"	art	noun/label="−subj"	adj
Declare	art/label="−declare"	noun/label="−subj"	verb	adj

table together with a general parser is used to parse device messages from the system controller to each device. A 'semantic function' maps the parsed message to a final action. With a machine specialist that can both understand and generate device messages, it is possible to intertranslate between multiple languages and hence devices.

Once a specialist has been built for each device, their interactions must be coordinated, taking into account sensory feedback from the environment. A cell language is defined to deal with the control of the cell, and a process plan is built for each machine in this language. As these process plans are executed the machine states are recorded in a cell state table, which is matched with a rule-set to determine the next cell action using rules of the type shown in *Table 10.6*. The system is event-driven in that the cell state is only checked when messages are received. CML assumes that messages can be transported correctly to the appropriate machine specialist or cell manager (e.g. using the technology described in Chapter 7).

Table 10.6 Examples of CML type control rules for a forging cell

(ACTIVE CELL) – (PERFORM LevelTwoRules)

[If the test to determine if the value of CELL is ACTIVE returns the value TRUE then the rule set LevelTwoRules is activated]

(LevelTwoRules
    (OPERATING MODE) – (PERFORM BatchRunRules)
    (MAINTENANCE MODE) – (PERFORM PreventitiveMaintenanceRules))

[These are the rules comprising the set LevelTwoRules. If the test of MODE=OPERATING returns true then the rules valid when the cell is operational are activated. Similarly for the value of MODE set to indicate that maintenance is required]

(AND (At Robot Loadstage) – ((Pass Rack BilletLocation To Robot)
    (Located Billet InRack)   (Acquire Billet InRack))
    (NOT (Gripped Billet))

[If the robot is at the loadstage AND a billet has been located in the rack AND the billet has NOT been gripped then the commands to pass the billet location to the robot and acquire the billet are executed]

Autonomous mobile robots

A frequently-encountered transport system for automated manufacturing is the Automated Guided Vehicle (AGV) that follows a wire set into the floor of a plant or some other form of fixed guidance. The AGV therefore has a restricted path. A number of groups are working on Autonomous Mobile Robots (AMRs) for the manufacturing environment. An AMR is intended to be able to navigate anywhere in the factory environment. There can be some confusion as to the definition of an AMR; some workers identify it as an industrial robot carried by an AGV. In this section, the AMR will be defined as an autonomous automated vehicle that is capable of navigation with three degrees of freedom in a two-dimensional floorspace.

The AMR concept has significant implications for the architecture of the factory. It will allow more flexible routing between processes, reduced free space clearance for manoeuvring allowing more compact factories, and most important, any changes in factory operation or layout will only lead to simple software changes.

Two of the most important enabling technologies for AMRs are the task-level programming of devices and the integration of the data from a number of different sensors.

There are essentially two forms of device programming language, implicit or task-level programming languages and explicit or device-level languages. Implicit or task-level programs, in which the device is not instructed explicitly, step-by-step, what to do, are at present solely implemented at a research level and then only to carry out simple tasks.

In task-level programming, the task, the job that the device is required to carry out, is described in a very abstract way. An instruction in a task-level program essentially represents a description of a state transition (the discrete step of changing from the current state to the next state) in the world of the device. Such a statement for a conditional navigation task is shown as *Table 10.7.* This statement is a description of an initial state of the device and of the desired state to be reached if the first state is encountered. It is assumed that the device controller has the necessary hardware and software to allow the high-level instruction to generate lower-level instructions to perform the actual state transition. How such transitions can be achieved will be discussed at length in the next section.

Table 10.7 A task-level statement to monitor a navigation task

---

$<$(LESSP (RANGE sonar$_i$) threshold$_i$)$><$(STOP MOTION)$>$

(IF the sonar sensor detects a range of less than the threshold value THEN stop the motion of the AMR)

The other key problem for the AMR is the perception and modelling of their environment. This environment will include fixed features such as walls, doors and stanchions. It will also include objects such as the manufacturing equipment itself, the position of which will vary infrequently with time. It will also include pallets, for example, the position of which varies frequently in time. The workspace will also include other AMRs going about their business. The AMR will have to perform a number of tasks such as transport, docking at workstations, recharging and perhaps manipulation.

Commercial products already exist that use laser scanning of fixed reflectors, odometry (distance travelled measurement) and a simple externally generated map of their environment to navigate around factories.

In order to carry out these tasks in a variable and dynamic environment, the most advanced AMRs are likely to have the ability to perceive and model their environment themselves. This will require on-board processing for navigation and motion control in addition to a connection to an overall ground-based planning system. This on-board processing is likely to take data from three further sources. The first source is a number of panoramic proximity sensors for object detection and local referencing for motion and position control (the usual solution for this is a number of ultrasonic emitter/receivers to provide fast short-range data). Secondly, a panoramic scanning laser range finder for long-range depth mapping of floor free space to allow position referencing and path control (faster to process and less dependent on external conditions than vision). There is also a need for incremental trajectory control, which can be obtained by optical encoders or odometry.

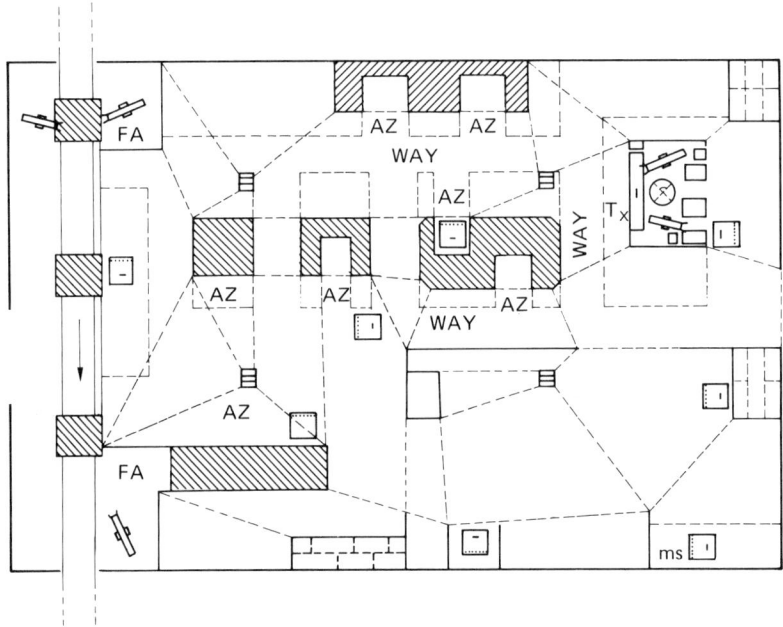

Figure 10.12 A cellular decomposition of a manufacturing workshop for an AMR (Ways = free areas, AZ = approach zones, FA = forbidden areas)

Multiple views of these sensors are then used to build a polygonal map of the obstacles in the environment of the AMR, which must take into account uncertainties in the position of the robot and sensor inaccuracies. This model identifies free space in the environment, which is divided into a topological model of free polygonal cells, as is shown in *Figure 10.12*. The connectivity between the cells is determined to give a graph describing paths through the environment. This graph can then be used to program the devices so that the AMR will have its state transitions described as transitions on the graph and then will find its docking position by processing sensor data, as indicated by *Table 10.7*.

The use of CAD models for intelligent machines

Within the subject of intelligent machines for manufacturing there are two important uses of geometric models, one for automatic path generation for an assembly robot and the second for model-based object recognition.

CAD generation of fine motions for assembly
Work in this subject assumes there will be CAD-based models of the objects that are to be assembled and of the assembly station. The starting point of the programming system is a description of a global plan by a human operator who describes the task as a series of localized operations or state transitions, the task level program. The computer then generates automatically these local 'fine motion' operations such as picking or insertion, which take place in three dimensions.

The CAD system stores the object geometrical data as a solid model and uses logical formulae to describe geometric constraints on the model. Shape features such as holes, shafts, bolts and open faces are used to describe characteristic features of the parts, and these shape features are used to describe the relation between features in an assembly. For example:

(featureOf A B)

states that feature A is a shape feature of part B and the assembled states are represented as primitive pairs such as:

inserted,
bolt fastened.

In this way we can describe parts and assembled states.

These can be used to generate the fine motions of the robot in three stages, the generation of an assembly sequence, the generation of robot paths for this sequence and the generation of grasping motions. Each of these stages is treated in a similar way with the generation of candidates and the filtering of these candidates by constraints until the required solution is generated.

In the assembly sequences, the assembly motions are constrained by the relationships between the parts. These motions are represented by vectors that represent the assembly or disassembly motion direction and its starting point. For example, the constrained motion of a vertical peg in a hole is such that its removal direction is on the positive z axis. To generate an assembly sequence, the disassembly sequence of a subassembly is reversed. The assembly sequence is then checked for collisions. This motion is then used directly to generate the robot motion, using a CAD robot model. The candidate motions are checked to ensure that the kinematic orientation and position of the end effector is possible, that the joint angles and motions of the robot are possible and that there are no collisions.

An example of an assembly task:

(assemble 'gear 2)

and the resulting description of the successfully implemented task-level plan is:

(inserted pin2 hole21 contact).

CAD model-based object recognition
Many groups of workers are studying the problem of recognizing and locating partially visible objects, for example, by using 3D models of the

objects to find them in range data collected from a vision system. Most groups have concentrated on the 'bin picking' problem, the location and retrieval of a jumble of relatively complex parts in a bin. These systems mostly work by locating a distinctive feature of the object, using the position of that feature to look for a second feature that verifies the first and then using the two features to predict a third that completely constrains the position and orientation of the object. With a number of objects, however, some of the predicted features may not be visible or the sensor may miss some of the features present so that an alternative strategy must be applied. This section describes a particular implementation of such a system described by Bolles and Horaud (1986).

The system is based on a CAD model of the part with four elements, an extended CAD model, a feature classification network, a planar patch (small flat surface element) model and a wire frame model. Each of these models is designed for a particular purpose, the feature network supports feature selection, the planar patch model is used to project range data for the object verification and configuration understanding, and the wire frame is used to display alternatives on the range data. To reduce the combinatorial explosion inherent in this sort of search, the search space is reduced by careful selection of the order of the features to be measured.

To extract the features from raw vision data, the discontinuities in the range data are detected and classified, they are linked into edge chains after artifacts such as shadows have been discarded. Each of these chains is projected onto a plane and segmented into lines and arcs. Then the surfaces adjacent to the edges are then analyzed, and this analysis is used to refine the edge data. This refined data of line and arc edges can then be matched to the line and arc edges expected from the CAD model, starting from the most distinctive feature, and hypothesizing and verifying the next distinctive feature. To check a hypothesis, the program predicts the range data expected, compares it with the real data and makes decisions based on the correlation of these. The predictions themselves are based on what the sensor would have seen if the object was in the hypothesized pose using the planar patch model.

Conclusion

This chapter has briefly reviewed what is possible with today's automation. It has also indicated what techniques may be applicable in the factory over the next five to ten years. Some of the tools developed as elements of these large demonstrator projects are likely to have significant effects on the design and operation of the automated factory.

Engineers are familiar with the problems in the real manufacturing world and should test the systems being developed against these. The manufacturing environment is a complex three-dimensional space that moves in real-time. Fast cycle times are required for economic production and irreversible damage to expensive mechanical hardware cannot be tolerated. The development of automated systems is a fast moving and volatile field that is driven by pragmatic manufacturing engineers and the available technologies.

Acknowledgements

The authors of this chapter would like to thank their colleagues, Paul Rogers, Jan Spence and Kevin Tilley, for their assistance and support in the preparation of the review material presented here.

Further reading

ALBUS, J. *et al.* (1981) *Theory and Practice of Hierarchical Control,* Proc. of 23rd IEEE Computer Soc. Int. Conf.

BATCHELOR, B. G., HILL, D. A. and HODGSON, D. C. (1985) *Automated Visual Inspection,* UK: IFS Publications

BOLLES, R. C. and HORAUD, P. (1986) 3DPO: a three-dimensional part orientation system, *Int. J. of Robotics Res.,* **5,** 6, 3–26

DUFFIE, N. A. *et al.* (1986) *Hierarchical and Nonhierarchical Cell Control with Dynamic Part-orientated Scheduling,* Proc. NAMRC-VIV, Minneapolis

FOX, B. R. and KEMPF, K. G. (1987) *Reasoning about Opportunistic Schedules,* IEEE Int. Conf. on Robotics and Automation, Raleigh, pp. 1876–1882

WECK, M. (1984) *The Handbook of Machine Tools,* New York: Wiley

WILLIAMS, D. J. (1988) *Manufacturing Systems, an Introduction to the Technologies,* Milton Keynes, UK: Open University Press. New York: Halstead Press

WRIGHT, P. K. and BOURNE, D. A. (1988) *Manufacturing Intelligence,* Reading, Mass.: Addison-Wesley

Chapter 11
Customer/supplier communication

The computer-based communication requirements for both large and small manufacturing companies are covered by the terms 'customer' and 'supplier'. In small companies, of course, most communication is in the role of customer or supplier. In large companies, the idea is frequently introduced that other departments should be regarded as customers, both as a commercial measure to ensure sensible costing, and to impose proper disciplines of quality and timeliness. Communication to neighbouring departments should not, of course, present great difficulty, but if the customer/supplier department is the other side of the country (or the world) computer incompatibilities within the large company can be as significant as for separate companies.

Incompatible systems within large manufacturing companies are quite common, since selection of computers has often been a local decision. It is easy, after the event, to assert that five or ten years ago people should have had the wisdom to anticipate the requirement to connect disparate computer systems together. Yet even now, with the writing clearly on the wall, not everyone finds it easy to contemplate a five-year strategy for CIM, and to take decisions based on that strategy.

This chapter will cover some applications of the CIM architecture components (as described in Chapter 7) that particularly affect the customer/supplier relationship:

- the use of networks for transmission of business data;
- the use of computer systems for electronic publishing;
- management of the physical distribution of products.

Customer/supplier networks

Transmission of business data is being revolutionized by the increasing liberalization of the PTTs (the national telecommunication authorities). Telephone networks can carry both voice and data on the same network; ISDN (Integrated Services Digital Network) is the integration of voice, text, image and data communication into a digital signal through a single high-capacity cable, based on CCITT recommendations. Sharing of lines by both voice and data offers increased flexibility and reduced network costs; optical fibre communication systems give vast increases in the volume of data that can be transmitted, as well as further cost reductions.

Connection can be made to networks from car telephones (attached, for instance, to portable PCs for transmission of business data). Private networks can be hired from the PTT; as discussed in Chapter 7, such networks, between different sites, are called wide area networks (WANs) to distinguish them from LANs (local area networks) within one site.

Customer/supplier communication based on integrated computer systems is intended to do more than save the cost of postage stamps. Major business objectives (as discussed in Chapter 2) are being pursued, including the reduction of product lead time and costs.

Business objectives of customer/supplier networks

The prime effect of these networks will be to transmit business information much more quickly. This can reduce lead times, or, by allowing more frequent communication, lead to a cheaper and/or better product. Immediate communication is clearly an important enabler for the implementation of just-in-time systems (as described in Chapter 9). Other benefits include avoidance of the cost of keying into the computer system printed documents that arrive through the conventional mail, and the consequent inevitable errors. All business data could be transmitted including:

- quotations and requests for quotations,
- CAD data, engineering analysis data, engineering change notices,
- orders, forecasts, order status and call-off data,
- capacity planning information,
- shared databases,
- process data, NC and robot programs,
- shipment documentation and invoices,
- product documentation, e.g. for printing at the customer's premises,
- periodic reports to suppliers on their performance in terms of product quality and on-time delivery,
- payment authorization,
- ordinary text messages, as an extension of the office internal system.

For all this data, security must be considered.

Security

Security can clearly become a major issue with so many authorized users of the company communications network and so many standard interfaces designed specifically to allow easy communication. The possibility must be considered that these users, and other unauthorized users, might access confidential data. Future product plans, competitive pricing data, customer lists, or salary levels might be exposed to someone with sufficient technical knowledge and determination.

In any one company, there will not be many people with the requisite skill and familiarity with the system software. However, if hundreds of other organizations can connect themselves to the company computer system as potential suppliers or customers, the system is much more exposed. If a computer is actually programmed to break a security system, it can make a lot of attempts very quickly (e.g. to discover a password by trying many possible combinations). The word 'virus' in this context is perhaps self-explanatory; it describes a program that can lie dormant until triggered by an event such as a date, or by being read, if it is a message. If the creator of the virus is being destructive rather than just showing off, the virus could then delete or corrupt data, while sending copies of itself to other users. Some very large networks have been 'infected' in this way.

There are, of course, many positive ways of trying to ensure security. System software has many checks built in and these have been well proven. For instance, all wrongly entered passwords can be recorded, so that by the time the 'hacker' has tried his first hundred passwords (with a few million still to go) the alarm bells will be ringing. There are rigorous checks to stop programs accessing unauthorized parts of computer memory. This discipline is, of course, important to stop a carelessly written program accidentally over-writing data being used by another program that is running simultaneously on the same computer. Security for databases include ranges of passwords (as discussed in Chapter 4), checks that the enquiry comes from the correct terminal, and controls over where the output is sent. Terminals can be restricted to keyboard input and display output (i.e. no printers or removable disks) so that people authorized to see confidential data cannot remove or copy it in any serious volume. Vital data can be kept on a separate tape or disk, and only loaded at the appropriate times (a technique used with payroll information).

One approach to maintaining security on customer/supplier networks is to have a single computer to interface with the outside world, allowing only rigorously checked messages to pass to the main computer network. A long string of machine code — intended perhaps to override the database security system — would be easily differentiated from conventional business data such as a request for quotation. Unauthorized access to the messages queued up on this interface computer, even if it could be achieved, would be much less serious than access to the main company network. Another precaution is to set up the system so that when anyone (i.e. a computer) telephones in, even with all the appropriate passwords, the receiving computer then phones back. This provides an extra check that the calling computer is who, or at least where, it says it is.

Precautions are also needed to secure what is termed an electronic signature, necessary if signed and numbered purchase orders, invoices and cheques are to be replaced by electronic messages. There must be no possibility that the recipient add another zero to the payment authorization. This will require at least a rigid procedure for copying, and retaining the copies of, all transmissions.

For really high security, encryption (as used for instance between banks and the cash dispending terminals) can be effectively unbreakable, i.e. a computer fast enough to try all possible combinations within a usefully short period of time is so expensive that the effort is not worth while.

Managed data networks

Managed data networks enable companies to subcontract the provision and management of networks. Such networks can link customers and suppliers (nationally and internationally) as well as different locations of the business. The cost of providing the skills for networking is avoided, giving a considerable saving in a complex and expensive subject. Other costs are reduced because telephone lines are being shared with other users of the network.

The user will be connected to the network with hardware and software specified by the supplier of the service. This may be proprietary equipment, which is not likely to matter to the user provided the service is reliable and cost effective. OSI and other standard protocols should also be available as regulations will differ across the world.

Apart from standards for actual transmission of data, there need to be standards for the data content. The network will be required to deliver the data accurately and promptly. In order to minimize the volume of data transmitted, the data needs to be compressed, eliminating, for example, blank lines and empty fields. The receiver of the data has, as a separate exercise, to convert it into a recognizable business document such as an order or an invoice. Standard protocols are needed for this data compression and the recreation of the original document on arrival, distinguishing carefully between part numbers and purchase order numbers, and between delivery and invoice addresses. The United Nations Trade Data Interchange standard (UNTDI) and the CCITT Office Document Architecture (ODA) have application in this area.

Another requirement is for the printing of formal documentation, such as required for customs, flexibly and remotely, including multiple-part copies with different data on different copies. Given the appropriate software and printer, such documents can be printed in total, with no need for preprinted forms because the printer also prints the boxes on the form, the pages being printed successively. EDIFACT (self-explanatory as the Electronic Data Interchange For Administration, Commerce and Transport) has developed from a desire to simplify the documentation and paperwork involved in international trade. It is not actually a protocol, but a hierarchical means of structuring data.

The supplier of the network will be responsible for providing a consistent level of service, measured by system availability and speed of response, among other factors. The flexibility of a large network to offer alternative paths through the network, as opposed to single leased lines on a smaller one-company network, can improve total system availability. The supplier of the network should also keep it technically up-to-date, so that new facilities and standards are quickly made available. Support for problem management is important; there needs to be a mechanism for reporting problems, and people with the appropriate skills available to tackle them.

The supplier will have in turn negotiated the use of facilities from the PTTs. Hence the economics of these services are dependent on the charging algorithms of the PTTs. If, instead of renting lines at a cost per line, allowing the suppliers of managed data networks to offer economies by sharing lines between their customers, the PTTs were to adjust the cost according to usage of the line, the economics of managed data networks would change substantially.

Value added networks (VANs)
Managed data networks can offer a specific service to a 'business community' group, for example automotive dealers or insurance brokers.

An enquiry for an insurance quotation could be circulated by a broker to insurance companies on a specialist network, and returned with a short list of the best quotations. The network has in fact 'added value' by doing more than just transmit a message. 'Armchair shopping', providing information about prices and availability, is a similar example.

Electronic message systems
Electronic mail over national and international PTT networks enables individuals to send and receive messages privately, under password control, from a home computer or via the company computer network. Messages are typed into a computer terminal (any common terminal is likely to be easily fitted with the appropriate hardware and software), temporarily stored on disk if required, and then transmitted to addresses which are marginally more complicated than post or zip codes. Thus electronic mail is functionally equivalent to company message-sending systems (discussed in Chapter 7) and can be linked directly to them. The benefits of the high speed are clear, and cost can be significantly lower than postal services, particularly if multiple copies are to be sent.

This type of service will allow everyone, not just people on large company networks, to have an easy and quick communication facility. Small companies can link to the networks, and with home computers becoming as common as television sets, it will incidentally become increasingly practical for people to work from home. In addition to having a 'help' service, available, the user would expect to be able to:

- create, update and transmit documents and messages,
- confirm delivery,
- hold distribution lists,
- store, find and forward documents,
- review and work on incoming messages,
- print documents and messages.

A form of electronic message transmission is provided by services such as the PRESTEL service of British Telecom. A central database is set up, and continually updated. Low-cost terminals, such as adapted television sets, can make enquiries and send simple transactions, usually with local call access. Travel agents making holiday bookings is the classical application of this type of system.

Standards for electronic message handling are supported under OSI by the CCITT X.400 recommendations, which are being widely adopted by the major suppliers. X.400 will cover both text and image processing.

FAX

FAX (Facsimile transmission) deserves a mention. It does not have technical glamour, but its quality improves with each development in photocopying technology. It will continue to be valuable for unstructured documents, for instance not within ODA or EDIFACT formats.

However, FAX is inadequate for transmitting 'live' documents that need to be developed further on the computer. It is not suitable for instance for CAD data if the receiver needs to view the drawing on his own workstations to develop tooling or NC programs, or to propose amendments to the drawing before returning it.

Transmission of CAD data

As described in Chapter 7, CAD data can be transferred from one system to another using available and developing standards. *Figure 7.5* illustrates the process. Implementation of these standards depends on the supplier of the CAD software providing the programs to unload and load the data; use of these techniques is continually developing. However, it is important to consider carefully what data actually needs to be transmitted. For instance, not every detail of a drawing has to be sent electronically for an NC program to be developed, or for a plastic moulding to be prototyped. Hence applications can be developed even though the current implementation of the interchange standards may lack some features. Many companies will use different interchange standards for different types of items, for instance electronic components, complex surfaces and mechanical parts.

Control of the exchange of CAD data will require a management process. A relational database approach would be required, able to flexibly control details about the information that has been passed to each supplier. Much of this data, for example about new products, future production volumes, failures of components in the field causing urgent rework, will be commercially very sensitive. The considerations discussed in Chapters 4 and 7 regarding the control of master data, which companies have copies of this data, and need to be advised if changes are made, are even more important in dealing with customers and suppliers than they are in internal communications. Patent applications, and the whole question of IPR (Intellectual Property Rights) will be affected by this easy mechanism for exchanging information.

Engineering change control

Effectively, the suppliers and even customers are being added into the engineering change process, a process that most companies find difficult enough to manage in-house. Depending on the type of business, suppliers and/or customers could need not only CAD data, but also bill of material data. The CAD system contains assembly and component drawings, the bill of material system contains the structure of each assembly and subassembly, including lists of components. Also in the bill of material system will be the details of those engineering changes that have been formally approved, together with the effectivity data, giving the date, batch number of other criteria by which the engineering change is to be implemented. Date effectivity is the simplest method, and if instructions are to be communicated to a supplier about an engineering change, then a date is certainly required.

For service requirements (where relevant) it will be necessary to record the history of engineering changes and previous service work. The service engineer could access the data on the network from the customers premises, in order to get exact status information. Electronic messages will be required as part of the engineering control system, requesting action, querying facts and providing information. Clearly, all this communication will be subject to security and confidentiality agreements, but the interchange of data in a CIM environment is going to be more complex than just the passing backwards and forwards of a file of neutral CAD data.

Application integration

Application software will have to be extended to accommodate the transmission of data electronically in a CIM system. For example, a purchasing system might be currently programmed to print purchase orders, and then expect on-line entry through a VDU of the typed or computer printed replies from the supplier. The process of sending purchase orders electronically will require extra control procedures, to monitor the orders as sent and ensure they are acknowledged. This extra control is likely also to be needed for the auditors. Receipt of quotations electronically will require programs that carefully edit and control the input, to trap incomplete or inaccurate data. The earlier on-line programs will still be needed so that any errors can be corrected. Interfaces into the financial systems of the company will also have to be changed to handle electronic invoices and payments. These control procedures are likely to include daily exception reports, since presumably the purchasing people will not want to be disturbed by error messages everytime a query arises, unless of course they have flagged an order as critical, so that they receive an immediate message if a query arises that might hold it up.

Other applications will need to have comparable modification. However, not all data will be transmitted electronically even in very advanced CIM systems. Another significant development in computing technique is In-House Publishing (IHP).

In-house publishing

Delivery of documentation for a technical product has been known to delay delivery of the product itself. Poorly illustrated or inaccurate documentation can spoil an otherwise good product in the eyes of the customer.

Prompt, well laid out proposals can, of course, be a great help towards winning a customer's business in the first place. If internal documents are improved and made more interesting, communication is more effective.

IHP is one of a number of self-descriptive acronyms (DTP, Desk Top Publishing is another) for what might be termed 'computer-integrated publishing'. The technical advances behind this perhaps unexpected application of computing are twofold: the developments in printing, particularly matrix and laser printers, and the availability of cheap processing power to handle the large volumes of data (as described in Chapter 4). IHP systems can run on all sizes of computers from centralized mainframes to distributed systems and personal computers. These systems have different capabilities; choice will be determined a number of factors, including print quality and volume, and whether colour is required.

A wide range of software is now available. Extensive use of WIMPs (Chapter 4) can make it easy to learn and use. Thus CAD drawings, and graphics displays such as charts and histograms, with parts lists from the bill of material database, can be inserted in text that has been prepared on a word processor, and the page displayed on the screen before printing. Photographs and freehand sketches can be read into the computer by scanners, and images can be created directly on the computer. Enormous choice of type styles is available, with 'intelligent' software to scale or rotate the constituent parts of the page as required. Industry standards are developing for specialist languages (Page Description Languages) to format the output. Camera ready copy can be produced.

Work stations with high-resolution displays will provide improved clarity of image for professional illustrators. Taking full advantage of the capabilities of an IHP system is likely to require both artistic talent and a sound training in computer technology. Image quality, such as line weights, texture and colour, graphic projections and font quality are all important factors in the finished product.

Distribution management

One of the earlier commercial applications of computers was for distribution management. In the 1950s a catering company in London actually built its own computers (the LEO series, the L standing for Lyons) to control office procedures and the stock levels of cream cakes in teashops. Technology has moved on, and for companies with a network of warehouses or retail outlets, CIM has very real application. Distribution management requires integrated information about stock levels, customer orders and expected arrival of new stock. Immediate decision is usually required. If stock can be accurately and quickly replenished then expensive shelf space can be saved, i.e. another just-in-time application. Customers can be lost because the stock is in the wrong warehouse, or because up-to-date information was not available about the next delivery or about alternative products which might have been offered.

Even though a distribution system apparently lacks some of the complexities of planning the manufacturing process, it still has 'interesting' scheduling problems. For instance, route planning for delivery lorries, planes and trains, with capacity constraints of weight and/or volume, can be a complex task. The need to monitor and control shelf life imposes further constraints. Rule-based AI systems are being applied to this type of problem.

Statistical forecasting, both extrinsic and intrinsic, is important for distribution businesses, particularly in the control of seasonal and fashion items.

Centralized and distributed data

Distributed databases, as discussed in Chapter 7, need to be considered. Each warehouse is likely to have some computing facility, for instance to enquire about stock levels and to print local paperwork such as dispatch notes, and will usually be responsible for its own stock, as a business control mechanism. Also a central function, either manufacturing or purchasing, depending on the business, will need to know total stock levels as well as those at individual warehouses. If the company has manufacturing facilities, these stock figures will be part of the MRP system.

There are advantages in holding customer order information centrally for credit checks, and even in a pure distribution business this may lead to stock levels being held centrally as well. However, if each warehouse has its own customers and there are a lot of local transactions that individually do not interest the central organization, then it may be a better business solution to give each warehouse its own database. The computing hardware at each warehouse could be terminals would be linked to a central computer in a star topology or local computers linked as a network. In terms of hardware, it is worth noting that a new warehouse can be built in less time than it takes to instal some computer systems!

Stock levels in warehouses are usually recorded in SKUs (Stock Keeping Units) rather than numbers of items (e.g. the number of bags of bolts rather than the number of bolts). The need for alternative 'units of measure', and to translate between them, is a specific feature required in the application software.

Hence, distribution management is another CIM application. It is now appropriate to look at techniques for implementing CIM systems in a controlled and positive fashion.

Implementation and the future

Chapter 12

Planning, implementing and managing CIM

The need for company-wide planning and implementation

The real long-term benefits of CIM can be achieved only through company-wide planning and implementation. Most parts of the company, to some extent, will be involved in the implementation process and most managers will be affected. Some managers will be sceptical about the benefits of CIM, while a few may even hope that the CIM plans will come to nothing and do their best to bring this about. To help forestall and overcome these problems, there are some simple guidelines that should be followed when planning the implementation of CIM.

First, the company must aim both at the short *and* long term benefits of CIM. CIM investment priorities should not be based only on demonstrated short-term benefits or savings. Certain programmes, such as setting up an engineering database, will yield their real benefits only in the long term, but may nevertheless deserve priority. In many companies, the real benefits increase exponentially with the number of engineering designs held on the system, so it can take several years before the full benefits of such a database are realized. However, these benefits can often be immense. Not only do they include cost savings in the Engineering Department, but, they also allow more designs to be considered and alternatives tried. Product designs and processes can be improved, which is the key to an improved competitive position and greater market share.

The second guideline is to try to involve all business functions right from the planning stage. Marketing and Sales should be aware that CIM can help a business compete more effectively or even change the basis of competition, and should have their say in identifying targets for improved performance. Since CIM does change the nature and volume of computing which takes place in the company, the Information Systems department should be involved in both planning and implementation. The best way to plan and implement CIM is usually through a multidisciplinary team, led by a CIM executive, and with the direct involvement of top management.

Thirdly, it is necessary to identify and concentrate on the CIM technologies most relevant to the company. Not all CIM technologies may be equally important for all companies, and there is a great risk, in terms of credibility and confidence in the eyes of other managers, if an attempt is made to implement all the technologies together. Planning means selecting and setting implementation priorities.

As described in Chapter 2, to achieve the real benefits available, CIM planning must begin with fundamental business questions. What is the basis of competition? In other words, what are the key product and customer service features? What must be done to compete effectively? Also, what are the critical criteria of our operating performance? The basis of competition may be price, quality, product features, delivery lead times or anything perceived as important by the customer. Once identified, the basis of competition can be translated into critical success factors. For example, a competitive price may derive from lower materials cost, or lower labour costs and so on. If the basis of competition is shorter lead times, this may result from more rapid product engineering and design, a shorter production cycle, increased inventory or improved distribution. Whatever the critical success factors are, they can in turn be defined in terms of explicit operating performance criteria.

Once these criteria have been identified, the performance of the company must be assessed against each of them. Ideally, these assessments should be analyzed and measured formally, but often this is not practicable and a skilled judgement must be relied on. The targets necessary for each of the criteria to bring operational performance up to the desired level must then be decided.

Those CIM activities which make the greatest contribution towards improving the operating criteria which underpin the businesses critical success factors will have an early position in the implementation plan.

Political sensitivity in implementation-planning

In setting the CIM implementation plan, it is important to take account of the political need to 'carry' company staff (both CIM enthusiasts and doubters) and maintain their commitment or, at least, interest during the implementation period. To achieve this it is important to include in the plan some activities which are both popular with managers and help them at an early stage. Whether these benefits are financial or in terms of some other measure, such as improved delivery lead time, it is important that they are both measurable, and relate to an aspect of the business that has previously been causing concern. A clear indication to staff at all levels, that CIM is working and justified, is valuable in providing a focal point for a steady build-up of enthusiasm for the whole implementation programme. Implementation of even a small element of CIM that works well is the easiest way of achieving this.

It is important to be sensitive to the political impact of a different mix of short- and long-term benefits. It requires more than just an analytical assessment to get the balance right, a balance that may differ from company to company. It must be treated seriously as a key political issue which will be an important element in a successful implementation plan.

The implementation phase

The implementation of CIM should be a coordinated top-down/bottom-up process. The CIM plan should be composed of a number of implementation programmes. Some of them will involve a single technical function or department; others will be broader, or even corporate-wide in scope. Each programme must be clearly defined, in terms of its nature, cost and benefits, responsibilities and time-frame. The preparation of the CIM plan is achieved through a dialogue between management and technical functions. This process will continue during the implementation phase.

The role of management in implementation is to ensure that all CIM projects operate successfully and achieve the full potential benefits of integration. This means, first, establishing the overall systems architecture for CIM and, second, setting up the mechanisms for corporate project management. In other words, management must define the technical and managerial framework within which the technical functions will implement CIM. The loop is closed with the usual submission of investment requests, vendor appraisals and progress reports, as implementation goes ahead.

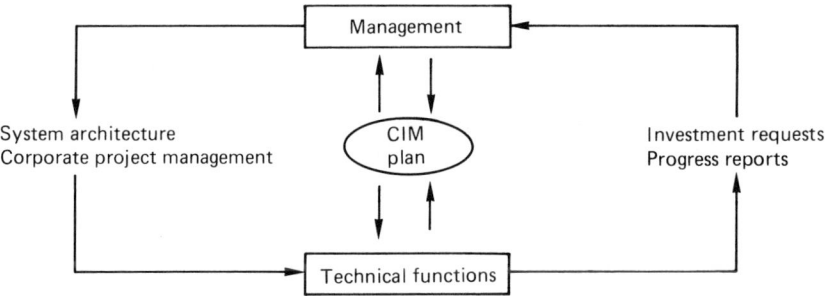

Figure 12.1 CIM implementation

Implementation (*Figure 12.1*) should start, or rather restart, with management setting up an overall systems architecture, since companies do not, typically, start from scratch. The first step consists of defining the technical framework (based on the functions described in Chapter 7) within which these systems will be operating. This framework will be needed to assess whether a given system, or a request for hardware or software, is compatible with the overall integration objective.

Develop a systems architecture

The systems architecture will typically specify the various systems to be implemented by functional or technical departments. It includes design and manufacturing engineering systems, which typically belong to the CAD and Group Technology areas, planning and control systems, including Manufacturing Resource Planning (MRP) and all its extensions, material handling systems, and productivity systems, which typically include CAM and robot systems.

The systems architecture will also specify a number of common elements, such as the nature and structure of databases, the communications infrastructure that will be used and standards for hardware and software. These standards are particularly important in avoiding a proliferation of incompatible equipment or systems.

In the course of setting up the corporate systems architecture, management will have to deal with a number of important issues, including defining the scope and degree of integration needed. In practice, this means determining which systems should be part of the corporate CIM system, and which should be stand-alone systems, a decision partly determined by the technology available and partly by operational needs. The former will have to be developed in accordance with corporate standards, in terms of databases, communications, hardware and software. Stand-alone systems, by definition, can be developed independently.

Conflicts may arise between IS people and users on the degree of integration. Ultimately, these conflicts can be resolved as to who should be in control of a given system. Users are often reluctant to have IS people interfere with their operations, or even to allow others access to their data.

Data accessibility and control is important. Integration means setting up an integrated database. This does not involve the creation of a huge single entity, where all the data pertaining to the company operations are stored but an integrated set of databases that appear to be one. Defining data accessibility and control means determining who should be allowed access to what part of the database. It also means identifying specifically who owns what data, i.e. who is ultimately responsible for the accuracy and updating of a given piece of the database.

Data security and integrity

As a company introduces CIM, it must address data security and the issues of data integrity. Data security involves: protecting data from examination or change by unauthorized persons, protecting data transmitted in telecommunications networks from loss or theft and protecting data integrity in case of system breakdown.

Well-proven techniques exist to cope with these, but there is a danger that companies newly entering the field do not recognize the importance and critical nature of security or the full consequences of breakdown. Management should be aware of, and willing to act firmly in the interests of data security.

The interfacing of the various CIM systems must be resolved, starting with the setting up of company-wide standards for hardware and software and the establishment of data communications networks. In many companies, one of the biggest obstacles to the rapid development of CIM, with potentially serious financial implications, is the difficulty of interfacing a wide variety of disparate and often incompatible systems that existed before real CIM planning started. In some companies there would be an advantage in starting again from scratch.

Users must have responsibility

Once the corporate systems architecture has been designed, the proper CIM implementation process can be initiated. It is important that responsibility for implementation should be assigned to the actual users of each of the systems or subsystems. For fairly complex projects, it may require the setting up of multidisciplinary task forces with various specialists, but in all cases, users should be represented or even chair these task forces.

Choosing between vendors

In many cases, the choice between different vendors is made during the initial planning phase. In others, considerable flexibility is left for a choice of vendors to be made during implementation. The choice is not as straightforward as it first appears. It should not be made merely on the technical performance of equipment; other important factors must be taken into account. For example, the availability of after-sales support can be crucial, and in some countries this will be the major determinant in the choice of supplier. The quality and range of that support must also be taken into account. Is after-sales support limited to hardware maintenance, or is a full range of software engineering and other services available from the supplier? Will the vendor tailor standard package programs to the company's particular needs, or will the company have to make the modifications using its own staff?

Clearly, the importance of this support will partly depend on the company's own internal resources to carry out these tasks. The right vendor decision will differ from company to company, even when installing similar systems. One of the techniques for choosing between vendors is shown in *Figure 12.2.*

Feature	Weight	Vendor 1	Vendor 2	....	....
Equipment price	4	5	3		
Delivery lead time	2	1	4		
Maintenance speed	5	4	2		
....	....	....	....		
....	....	....	....		
....	....	....	....		
Software support	5	5	3		
Total	18	....	....		

Figure 12.2 Choosing between vendors

Vendor performance is listed down the left-hand side under various headings. These may include, for example, equipment price or implementation cost, delivery lead time, likely time-lag in getting maintenance support to the company location, availability and quality of software engineering at the vendor, availability of other compatible equipment which may help later to extend your CIM systems, availability of back-up services or computing facilities should the company's equipment fail, and so on. The competing vendors are written across the top of the page at the head of each column.

The implementation team must then consider each aspect of vendor performance in turn, take into account the facts relevant to it, and give each vendor a score between, say, 1 and 5. Once this has been done, the implementation team must discuss the company's requirements in each performance area and, in the light of this, give a weight to each of the performance headings. If in-house resources in areas like software engineering are very good, then this can be given low-weighting. Once weights have been allocated, the method is to multiply each weighting factor by the relevant score given to each vendor and then total for each vendor across all performance factors. A comparison of the results provides an initial ranking for the vendors, and a sound basis for comparing and choosing between them.

Implementation should proceed step-by-step

Management should discourage users from trying to embrace too much too fast, as they are often tempted to do. Solid progress is achieved more rapidly by taking a step-by-step approach, enabling management to ensure that the staff are properly educated and trained before and while the CIM plan is implemented, that the actual costs linked with CIM implementation are properly controlled, and that benefits expected from a system are validated, even on a small scale, before the system is extended. Each user or user group should move judiciously and document one system at a time. Ultimately, it will be quicker than moving fast and getting bogged down in numerous implementation difficulties. Documentation of existing and planned CIM applications is particularly critical and something that users are often tempted to overlook. Although users must remain the driving force, the use of specialists to monitor documentation progress and standard is a valuable additional check.

Project control

It is advisable to monitor implementation progress using network planning methods. This is important not just to achieve implementation in the shortest possible time, but to ensure that progress, use of resources and benefits occur in line with plan and that corrective action is taken when necessary. This action may involve adjustment of the original plans, which is fully acceptable provided it can be achieved within the planned resources and without loss of benefit. It is vital to efficient implementation that plans can be readjusted. There are two reasons for this. Firstly, CIM implementation plans tend to be broad and technically complex. They often include multifaceted projects, subject to unforeseen circumstances and obstacles, so that some changes in the plan will be inevitable. Secondly, remember also that CIM technology may be changing faster than the implementation program itself. Management may have to reconsider technical options from time to time. The overall CIM plan itself should be viewed as a master plan and not as a hard and fast plan specifying all details.

Do not re-invent the wheel

In implementing CIM, there is an advantage in relying on commercially available software or services. Do not assume that apparent operating differences between companies mean that new systems must be tailor-made from scratch for each. This is particularly relevant for software development. Companies often prefer to develop their own software because they believe commercial packages will not fit their special needs. But such differences and special needs are almost always exaggerated. Management often does not realize the price it pays for this, mainly in terms of cost and time, but also in terms of less reliability, lower programming standards, poorer documentation, limited maintenance and enhancement potential, as well as limited educational support. In all these areas, commercial packages far out-perform packages developed in-house.

Many companies started developing their own MRP package some ten or more years ago because no really good commercial package was available to meet their requirements at that time. Millions have been invested in such developments. In many cases, even after so much time and money have been spent, management finds itself with a mediocre package. Many companies would be better off scrapping their system and starting afresh with a commercial package, painful though this decision may be.

The need to manage the impact on people and organizations

Employment levels

So far this chapter has dealt with the technical and business aspects of implementing CIM. There are, however, many other factors to take into consideration.

The significant impact of CIM on people and organizations must be anticipated and managed. There is a wide range of social, organizational and managerial issues linked with engineering and manufacturing automation.

When managers ask the question: what about the people? they obviously think, first, about the impact of CIM on overall employment levels (*Figure 12.3*). Automation cuts jobs where it is applied. Yet CIM users know that the impact of CIM technologies on employment levels is not all negative.

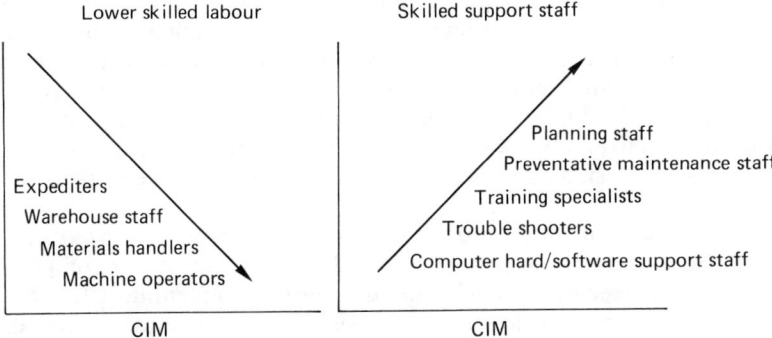

Figure 12.3 Employment level

Typically, a number of lower-skilled jobs will disappear, like low-level expediters, warehouse attendants, materials handlers and machine operators. These job losses weigh heavily, first in terms of sheer numbers, but also, given their social implications, in terms of newspaper headlines and social anxiety. CIM will also, typically, create new jobs, and these will tend to be more highly-skilled jobs, like planning staff, preventative maintenance staff, training specialists, trouble-shooters, computer hardware and software specialists and support staff. This will involve hiring new people and redeploying staff within the company, involving a lot of education and training.

It is important to ensure that this change in jobs brings about job enrichment rather than the reverse. There have been companies who have moved highly-skilled machine operators to lower-skilled and monotonous monitoring jobs, after their CNC machine is incorporated into a DNC or FMS system, foremen to the job of skilled operators who, in turn, are downgraded into lower-skilled jobs. Understandably, this has a strong demotivating influence and creates resistance to further automation. It is something that must be avoided.

In summary, there are two important impacts of CIM on the skill pyramid. Firstly, the number of engineers, technicians and foremen may increase. They will spend more time on managing and the skilled aspects of their job and less on mundane tasks such as cumbersome, repetitive calculation or expediting. Secondly, the number of lower-skilled operators and unskilled staff will decline.

Preparing people for CIM

The reaction of people to changes in working practice that go with CIM will be different. In most cases, the acceptance of CIM is greatly improved if the company makes the effort to prepare its people for this new way of working. Japanese experience is useful here. For example, typically Japanese management cleverly introduced robot developments into unpleasant work areas first in order to get them accepted.

Training
To build motivation for CIM, staff (from top to bottom) must be provided with ample information, education and training in CIM. More formal courses for CIM are becoming available, though much of it covers the individual CIM technologies rather than how to deal with the complex problems of implementing integration of the technologies into the business. Education and training are also needed to build up a general awareness of CIM throughout the company, and to highlight the extensive benefits that can be obtained from integrating operations, even to staff in functions or departments which may not be directly concerned by the implementation effort. Training is also required to upgrade the skills of the various people directly involved.

New academic initiatives are developing to satisfy this need; for instance, The CIM Institute at Cranfield, UK, offers courses for company directors, senior executives, engineers and staff implementing CIM that cover both the broad business issues as well as the technical aspects of CIM.

It is also important to ensure that career paths are readjusted to ensure that CIM is viewed as a source of opportunities rather than as a threat and that where necessary, additional incentives or rewards are provided to offset initial resistance or compensate for changes.

To build up an awareness of CIM in the corporation, the whole CIM development plan, i.e., its 'raison d'etre', its objectives and its implications, must be communicated to all levels of the company. Ultimately, everybody from senior management to supervisory levels and employee representatives must understand and appreciate what CIM means to the company.

Interaction of technical functions

CIM will affect the organization of the company; in particular there will be a change in the interactions of technical functions. Product engineering, manufacturing engineering, manufacturing and quality assurance, start working together more closely. The immediate availability of information, which the computer provides, facilitates more effective communication between functional departments.

The whole information processing area is also affected, since CIM vastly expands the amount of information processing taking place in the corporation. It also increases the degree of interdependence between various types of computing. As a consequence, the role and organization of the IS function may need to be redefined. Care must be taken in organizing the interactions between technical and business computing. Business computing people must be well-prepared and trained to handle the real-time demands of technical computing.

Senior management responsibility for CIM

The ability to implement successfully, or 'to get things done', is more important than any other. It demands the ability both to plan formally and to foresee likely practical difficulties before they arise. The person charged with implementation must be able to control against formal planning networks and know what action to take to bring a programme back to plan. Above all that person must have the grit and determination to overcome obstacles as they arise and not be easily discouraged.

Given its implications, the job of planning and managing CIM should be a senior management responsibility. It is not a minor task than can be done by an existing functional manager, for he will not be able to bring to it sufficient breadth of experience to give adequate weight to all the functions involved in integration. Once someone responsible for CIM planning and management has been appointed, all functions must be involved in the process. CIM must remain close to the actual user.

A big staff is unnecessary. The mission is to orchestrate, stimulate and manage the effort, in direct cooperation with the company's functional or technical departments.

Stages of CIM development

To help assess where a company stands in relation to CIM, the typical stages of CIM development can be characterized (*Figure 12.4*):

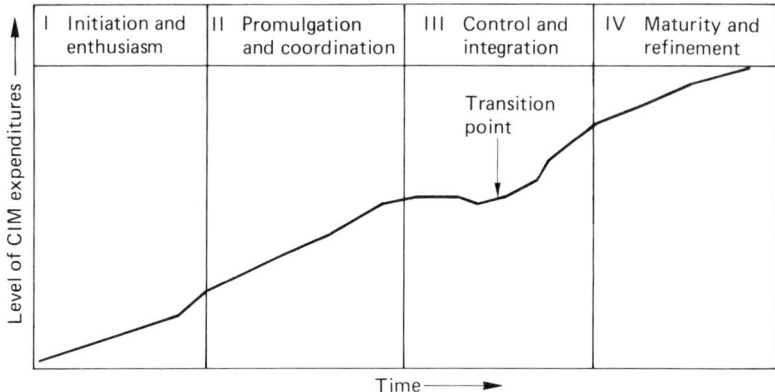

Figure 12.4 Stages of CIM development: expenditure

Stage I   Most companies start with a phase of initiation and enthusiasm. One or several CIM champions emerge. The amount of expenditure is still small, but management debate is intense.

Stage II   Once the go-ahead for CIM has been given (implicitly by a sort of 'laisser-faire', or explicitly through the process described here) the promulgation (and hopefully coordination) stage starts. The level of CIM expenditure increases dramatically, and everybody is busy implementing various systems.

Stage III   This proliferation phase is often followed by a management reassessment of the situation. Excesses have to be corrected, and a certain degree of discipline is introduced. CIM expenditures go through a holding phase, while consolidation takes place. It is the stage of control and integration.

Stage IV   Integration becomes a reality. This is the stage of CIM maturity and refinement. Very few companies have yet reached this stage.

Several indicators can be used to assess a company's stage of CIM development (*Table 12.1*). For example, the initial driving force may be individual initiative. Then, management initiative takes over. In Stage III, management controls become dominant. Finally, the system has gained its own momentum.

Another indicator is a growth in CIM expenditure. It starts off slowly, then it accelerates. In Stage III, it is often redirected. Finally, the company reaches a steady balanced growth.

In Stage I, CIM applications are limited to a few islands of automation. Then, typically, CAD/CAM links develop. Integration with other systems is reached in Stage III. The full range of CIM applications is exploited in Stage IV.

Table 12.1 Stages of CIM development, indicators

Indicators	I Initiation & enthusiasm	II Promulgation & Coordination	III Control & integration	IV Maturity & refinement
Driving force	Individual initiative	Managment initiative	Management controls	System's momentum
Growth	Patchy	Accelerated	Redirected	Balanced
Applications	Islands of automation	Extensive CAD/CAM links	Integration with other systems	Full range of CIM applications
Equipment types	Proliferation of equipment	Some limitations	Strict standards	Standards less relevant
Interface with IS	Little interface	Some sharing (data & resources)	Integration with DP data bases	Common distribution network

Stage I sees a proliferation of equipment types. Some limitations are introduced in Stage II, often followed by strict standards. In the maturity stage, standards become less relevant, because they are built into the system.

The last indicator deals with the nature and degree of interface with IS. In the beginning, it is largely independent. Then, there is some sharing of data and resources. Stage III sees the beginning of database integration. In Stage IV, the company operates a common distributed network. At the end of Stage IV, computer integrated manufacturing has become computer integrated business.

Potential pitfalls

Finally, there are many potential pitfalls in the area of CIM planning and implementation that must be avoided. The first is to ignore CIM or underestimate its impact. This kind of management myopia is frequently found in conventional industry sectors, or in medium-sized companies. Management states: 'CIM is not for us'. In most cases, the surprise will be cruel after a few years. The second pitfall is to lose touch with the rapid pace of CIM developments. What was true two years ago about the availability, cost and performance of a given system, may be totally outdated today.

There is often a temptation to postpone all investments until a given technology has matured or stabilized. This is the classical 'Let's not hurry, we'll catch up later' response. This reaction is based on a misconception about learning curves. There is no such thing as leap frogging in the CIM area. It is essential to build on individual experiences.

CIM opponents or sceptics must not be allowed to block or slow down the adoption process. Opposition to CIM can come from the Financial Controller, worried about immediate paybacks, or from IS, worried about losing its monopoly. It also often comes from managers, even in the technical area, who fear that the new technologies will reveal their inadequacies, or limit their power.

Finally, horror stories about advanced manufacturing technology in other companies should not be allowed to discourage CIM implementation. Failures in implementing CIM are almost always linked to poor planning, insufficient training or inadequate management supervision. Management should look at companies, particularly competitors, who have successfully implemented CIM and learn from them. Of course, expectations of CIM must be reasonable. CIM cannot solve all management problems. Neither is it a panacea for poor management.

CIM can provide major advantages, help improve the general productivity, efficiency and effectiveness of operations, and give a significant competitive advantage. It can also produce a sound and flexible manufacturing business that can look with confidence at the challenges and opportunities of the years ahead.

Chapter 13

Continuing education and the future of CIM

But more advanc'd, behold with strange surprise
New distant scenes of endless science rise . . .
Th'increasing prospect tires our wandering eyes.
Hills peep o'er hills, and Alps on Alps arise!

Alexander Pope (1688–1744)

The two subjects in the title of this chapter are intimately connected. Readers of this book should acquire sufficient knowledge to work on a CIM strategy, and to implement appropriate CIM systems, taking particular care to simplify the business systems before applying CIM technology. However, as Pope evidently foretold, CIM will continue to develop. It is important to keep pace with these developments, and this will require a continuing education exercise. The future of CIM will depend on two factors in particular: how CIM is implemented on a pragmatic basis in manufacturing industry, and the work which goes into developing new techniques.

It is necessary therefore to understand something of the national and international programmes that support CIM; industry, governments and academia are working together with a concentration that suggests that CIM will be around for a long time.

Continuing education in CIM

It is of course the fantasy of those involved in each new technology that this is the one that is absolutely vital to the business; attend a few courses and the key to the executive washroom will inevitably follow. Nevertheless, it does seem, even in the cold light of day, that the type and scale of decisions that are required for CIM would make some knowledge of CIM essential to most senior people in a manufacturing company, and those aspiring to reach such positions.

Using this book to keep the subject in perspective, a selection can be made from the many articles on CIM and its component subjects. Time constraints will prevent more than a percentage of this being read carefully, but such articles can be more immediately up-to-date than a textbook, although sometimes too specific. Equally there is a profusion of conferences and exhibitions. Selection is again a problem, because of both time and cost. One major industry exhibition a year would seem to be a minimum to keep up-to-date, and probably sufficient for most people. The contents of this book have hopefully guided the necessary selection processes. To be able to judge what not to read, and which conferences not to attend, is important.

Hands-on practice

Knowledge of the computer technical subjects can be substantially improved by hands-on practice. This applies for instance to CAD systems, expert systems, AI and other languages, word processors, windowing systems, database and production control systems, including the logic of MRP. Packages for these functions can run on PCs, so should be available at educational establishments and possibly affordable for education at home (cost justified on the basis of improved future earnings!). However, some subjects, e.g. NC and robotic programming, and the capabilities of high function work stations, clearly require more expensive equipment.

It does not take a great deal of time to acquire basic experience of these individual subjects as components of a CIM system; on average one to two days each should be enough. However, to go much beyond this level, significantly more time is required. Some months are needed to become a real expert at a CAD system, for instance. More extensive hands-on practice in a subject often requires a choice between different commercial systems. The initial familiarity is reasonably generic, but as more detail is learnt, the knowledge becomes increasingly specific to a particular equipment or implementation. Ordinary product education can contribute to a CIM education programme, with the advantages of being fully up-to-date in the detail and of immediate use.

Formal education courses

Some readers of this book will already be on formal education courses. Others may be interested in such a course, so that they can acquire a deeper education in CIM that can be obtained from private study. One alternative is to pursue a Masters course in CIM or the closely related subjects such as Manufacturing Systems Engineering, Advanced Manufacturing Technology or Computer-Aided Engineering. Such courses typically take one year, or the equivalent part time, and includes both lectures and hands-on project work. The approach on such courses is vocational and pragmatic. Potential students do not need to fear a return to the highly theoretical exercises of undergraduate days. Project work will by definition be industrially relevant, and can often be oriented to the special interest of each individual. Many such courses have been substantially funded by industry.

Of course, the majority of people cannot contemplate a year away from work to pursue their studies, however profitable it might be in the longer term. Hence, many universities and professional institutions identify shorter courses, either as a measure of serious study or to be accumulated to an academic qualification. For example, the American Production and

Inventory Control Society, fully supported by its British counterpart, runs a Certification Programme targeted to ensuring that the members know how to get the best out of the tools of their profession.

Distance learning

Distance learning is increasingly used as a means of study. The Open University in the UK offers courses in Manufacturing Systems. The PACE project in Europe is developing a satellite network to deliver advanced continuing education in six subject areas, including advanced manufacturing and expert systems. US universities have for many years offered remote education using TV to broadcast lectures, with telephone lines for live questions from the audience. World-wide satellite broadcasts (for instance in 1987 on expert system technology) are becoming familiar.

Extent of education needed

The question of how much CIM education is appropriate for people in different roles in a manufacturing company is important both for individuals and for companies implementing CIM. For the individual, the analysis will depend on the career path in mind; for the company an education programme is needed.

In-company education

A company's approach to CIM education will probably reflect the organizational style of the company. One proposal might be, for instance, that only a small number of people in the IS department need any real understanding of CIM. They will then provide the CIM systems for the rest of the company. CIM would be seen therefore as just a set of additional features to an existing system. For example, if the stores people are to see the stock levels of a company's suppliers, advanced networking technology, complex legal negotiation and man-centuries of work on standards committees may be required, but the storeman can learn to use the new system effectively in a few minutes.

However, this approach precludes the majority of the people in the company making much of a contribution to analyzing their own business problems (along the lines described in Chapter 2). If, for example, the purchasing people have not been given some reasonable education in the possibilities of networks to the suppliers, they cannot make a knowledgeable and confident business case for them. As a consequence, they may well waste time and effort on less profitable improvements to which they relate (perhaps giving preference to suppliers who are geographically closer, to whom drawings can be sent by courier).

As a comparison, when MRP was introduced it was held that perhaps 50% to 70% of the people working in the company needed to receive some degree of education and/or training in the subject. Education in CIM will be eased by the MRP and material logistics education that will usually have preceded it. This is because many people in manufacturing industry have now been introduced to the computer as a tool, and also to the associated needs for accuracy and discipline, significant new concepts when first introduced.

What is different about CIM is the breadth of knowledge which will be required of the people who will use it (MRP plus CAD plus database plus networking etc.). Hence education in CIM needs to be spread widely, varied to the level of people's interest, and would be based on the ideas in this book. The programme would need to include:

- senior executive overview of the development of a business case for CIM (costs, benefits and probable timescales) and the mechanisms for implementation,
- longer sessions for middle managers, with much of the same general introduction, and then more focus on specific areas of responsibility and relevant systems integration required,
- more detailed discussion of specific aspects for junior managers and professionals.

Appropriate strategic decisions can simplify the education requirements. If MAP is selected for shop-floor networking, then expertise in alternative networking technologies is not required (this is, of course, one of the benefits of standards). A general decision about which terminals are to be used can save much debate about which combination of processor, VDU, disk and printer is appropriate for each individual position. (A decision to do nothing about CIM, and to implement no new computer systems, can simplify matters even further, of course — if only for a short time.)

These and many other decisions relating to CIM strategy are important and will affect the company for many years. Such decisions should be taken in the boardroom by directors who have some understanding of the subject; the people recommending the decisions to the directors will need to understand the subject at a greater depth, and relate it to how the business works. However, education, either for a company or for an individual, is a long-term investment. It is important therefore to look where CIM is going in the future.

The future of CIM

The move toward integration is an evolution of earlier computer systems, firstly batch and then on-line. Inevitably the question arises, 'What comes next?' Both individuals and companies want to be sure that time and money invested in CIM will not suddenly be made obsolete by a newer technology.

It is clear that computers will continue to increase in power and reduce in cost (see Chapter 4), enabling more and more processing of data to be economically performed. The answer to the question 'What comes next?' in technical terms may well be that the next development in the application of computing technology to manufacturing industry will be the widespread use of AI techniques. AI and expert systems were described as important components of CIM in Chapter 7, in the context that their use is not yet common (as discussed in Chapters 8, 9 and 10). However, AI is the basis of the widely publicized Japanese Fifth Generation Project.

Nevertheless, if AI becomes such a large component of CIM that computer systems in manufacturing companies become 'AI' systems, rather than 'CIM' systems, the CIM systems will still be the basis on which the AI systems are developed. Data for AI systems will need to be extracted from the CIM databases; rules for AI systems will be developed from the attempts of existing CIM systems to solve the problems. Hence, knowledge of CIM as it has been described in this book will be fundamental. Chapter 1 suggested a possible first law of CIM, that however clever (or intelligent) a system, it will be improved by someone who understands in depth both the system and the problem it is trying to tackle. This is likely to still apply if AI is extensively used in manufacturing industry.

On a different dimension, CIM may be extended from Computer Integrated Manufacturing to Computer Integrated Management. There are other situations than manufacturing industry where stock levels, resources and conflicting demands have to be carefully scheduled, for example hospitals and many local government functions; as facilities become more automated and computerized, planning and control will be increasingly important.

Commitment to CIM development

There is wide national and international commitment to the future of CIM techniques; CIM is not just a 'flavour of the month', any more than were on-line systems.

The annual LEAD Awards (Leadership and Excellence in the Application and Development of CIM) by CASA/SME (the Computer and Automated Systems Association of SME) are an instance of a strategic industrial commitment to CIM. Major exhibitions are devoted to CIM or include CIM demonstrations as central showpieces. MAP in particular attracts a lot of attention in international exhibitions. MAP is important because the manufacturing shop floor is so central to the development of CIM, and because, as described in Chapter 7, it is an implementation of OSI. As early as 1986 the UK Government sponsored an exhibition called CIMAP, which emphasized the significance of MAP as a component of CIM.

There are many other major projects that are relevant to CIM projects. The ICAM studies by the US Air Force (discussed in Chapter 3), programmes by CAM-I (Computer Aided Manufacturing International Inc.), the US National Bureau of Standards work on IGES, the NBS Automated Manufacturing Research Facility, the Japanese Flexible Manufacturing System with Lasers at Tsukuba and the FAMOS programme within the structure of Eureka in Europe are just a few examples. It would be hard to find a large or medium-sized manufacturing company which is not participating in such projects; they represent a commitment, if not always to CIM as an acronym, then to the advanced technologies which are the components of CIM. Perhaps the largest such project at the multinational Government level is ESPRIT.

The ESPRIT project
The European Strategic Programme for Research and Development in Information Technology started on a large-scale five-year programme of work in 1984 with a budget jointly funded by the European Community and about 200 industrial companies. Funding for the follow-on phase, ESPRIT II, was agreed in 1987. CIM, one of five major work areas, was divided into subareas:

- Integrated system architecture,
- Computer-aided design and engineering,
- Computer-aided manufacturing,
- Machine control systems,
- CIM systems applications.

The architecture project has been assigned to a group of nineteen major European companies and is known as AMICE (European CIM Architecture, in reverse). The architecture itself is known as CIM-OSA (CIM Open Systems Architecture), which the participating companies are committed to implement.

CIM-OSA develops a model of the enterprise from two viewpoints, the business user and the equivalent manufacturing technology. It is a completely general model, with the two viewpoints reflected in two different models — the enterprise model and the implementation model. Particular models are derived by instantiation of a generalized reference model.

Other ESPRIT CIM projects are related to the CIM-OSA architecture, including the CNMA project (Communications Network for Manufacturing Applications). Eleven European companies are involved, including users, vendors and an academic partner. The third phase of CNMA is developing Conformance Testing Tools in accordance with MAP 3.0 specifications, indication of the part which standards will play within CIM.

International standards
Future developments in international standards will be very important to CIM. Some standards and corresponding functions are specific to manufacturing industry, but there will be cost advantages in using more general standards where possible. As described in Chapter 7, the processes for standards development are complex; there are risks in moving too quickly, just as there is an obvious cost in standards not being available as soon as possible. The specialist knowledge to understand the extensive documentation which goes into standards definition is rare and expensive, and most of it belongs to companies with direct commercial interests.

As discussed in Chapters 1 and 2, the commercial benefits of all aspects of CIM, including therefore investment in standards, must continually be evaluated. It seems unarguable that there is a benefit to society on a world scale from cheaper manufactured goods. However, costs will actually increase if each manufacturer has to write complex software to introduce each change in production method — and to pay highly for specialist hardware and software skills. From the perspective of an individual nation or company, life will be much simpler if connections to customers, suppliers and shop-floor equipment are easy to specify and instal.

Hence the reduced costs which will result from (effective) standards is generally beneficial. However, the standards relevant to CIM will be continually developing for the next few years, so it is not an adequate strategy to wait for them to be completed; progress must to some degree be monitored and understood, or else expensive mistakes might be made. This is clearly accepted by the many companies who participate, either as suppliers or users, in standards groups.

The European Workshop on Open Systems (EWOS) is an example of how these groups are developing. Its first Technical Assembly was held in 1988, when its members were:

CEN	Comité Européen de Normalisation
CENELEC	Comité Européen de Normalisation Electrotechnique
COSINE	Cooperation for Open Systems Interconnection Networking in Europe
ECMA	European Computer Manufacturers Association
EMUG	European MAP Users Group
OSITOP	Open Systems Interconnection Technical and Office Protocol
RARE	Réseaux Associés pour la Recherche Européenne
SPAG	Standards Promotion and Application Group

Offical standards bodies, manufacturers' groups, users' groups and research bodies are combined in an attempt to ensure that standards as they develop are universally acceptable. All of these groups have other direct relationships with each other and also with worldwide groups. For example, EMUG and OSITOP have common working groups and also are members of the World Federation of MAP/TOP User Groups, which includes User Groups from the US, Canada, Australia and Japan. Similarly, CEN/CENELEC have continual contact with ANSI and the national standards bodies. Furthermore, the individuals who compose the committees which prepare and approve the documents (the objective of the whole exercise) are a relatively small band of specialists. A group of people might meet as one committee, then again as another committee with minor change of personnel and with people having different roles, e.g. an individual could represent his company on the first committee and then a user group or standards body on the second.

It is of course easy to assert that this is yet another example of bureaucratic over-kill, but it is not; the problem is real. If standards are not seen to be based on extensive discussion, if they are technically obsolete or if they are simply badly drawn up so that they do not work, they will not be used.

The intention to eliminate technical barriers from Europe for the 1992 deadline will have a major effect on standards development, another reason why it is not possible to forecast accurately the timescale of results from this world-wide standards definition process. European standards (acronym EN, plus HD for harmonization documents and ENV for initial standards) relate to the national standards they replace or supplement, and also to world markets – an interesting political and intellectual challenge.

Academic/industrial cooperation

Many universities and colleges are installing equipment and setting up institutes for education and research in CIM, which is becoming a valuable source of academic income. Government agencies are fostering this association, for instance, the Science and Engineering Research Council in the UK and the National Science Foundation in the USA. One key objective of this cooperation is to ensure that full advantage is taken of academic research projects. Certain types of problem will be solved in industry before the university research grant could be approved, whilst other problems are sufficiently general as to be very suitable for the university research environment. A particularly difficulty in this environment can be the ownership of research results, e.g. products or patents.

Some of the equipment used in these programmes is donated by the manufacturers; cynics can (and do) point to the direct interest which many of these donors have in earning the students' good opinion. However, it is hard to see that students who sit for a few hours in front of a terminal will have been hypnotized by the badge on it, and will henceforth buy only that brand of computers. Also, commercial software may be unsuitable for student requirements because it explains too little or too much. Hence equipment donations to universities do reflect a genuine concern that education in the latest technologies is important for students who are about to enter industry. Hopefully one result of this cooperation will be to increase the attractions of manufacturing industry to students.

Government-funded projects in CIM and its constituent advanced manufacturing techniques are part of national attempts to move toward what has been termed a 'brain state'. The alternative to training engineers to implement CIM and related technologies may be no manufacturing industry, or a manufacturing industry which is an offshore assembly operation, existing only to evade tariff barriers. Most countries will prefer the first option; the wealth generation and employment provided by manufacturing industry is desperately needed.

This book is itself the product of cooperation between academics with current business experience and people working in industry with real involvement in teaching.

Summary

CIM is real. It takes advantage of a logical evolution in computing technology to improve the products and profit of manufacturing companies. It is changing all the roles in a manufacturing company. For some companies, their introduction to CIM will be the networking of data to customers or suppliers. Others will see increased electronic communication within the factory, particularly the closer integration of design and manufacture, as the way forward. CIM will itself evolve as integration of computers becomes commonplace, and the standards become stabilized, probably toward increasing use of AI and expert systems. In all situations, CIM will be more than just a new technology.

CIM requires important and expensive decisions about strategy in all aspects of the company business. Because of the importance of manufacturing industry to national wealth, CIM is a factor in national and international politics. It affects universities and technical colleges in terms of the educational challenge it poses.

The book has endeavoured to put the subject of CIM into perspective. It has described the tools available so that managers and professionals in industry, as well as postgraduate students, can use them to simplify business systems and increase profitability.

Tutorials

A good measure of whether a subject has been understood is the ability to view and present it from a different perspective. The tutorial exercises in this section are intended to help the student take this different perspective, and to evaluate whether he or she feels confident with it. There are of course no 'correct' answers.

Chapter 2   Business perspectives for CIM

(2.1)   List the business objectives of the company you work for. If you are a full-time student, answer the question for a manufacturing company you have worked for, and then for the university or college you are attending.

(2.2)   Specify the main products of the company you work for (or a previous company), and the market share of these products. How are these likely to change over the next five years?

(2.3)   What are the main features of your company products which differentiate them from competition?

(2.4)   Consider a business problem at your place of work, one that personally causes you grief. List the people and departments who contribute to, or suffer from, that problem. For each entry on this list, identify what you think will be their attitude to the problem, and how important the matter is to them. From this, can you define a business case for solving the problem that takes into account the concerns of all the interested parties?

(2.5)   Consider the computer terminal you most recently used (business, not privately owned) and evaluate how it (and/or the system to which it was attached) may have been cost justified, who might have done the calculations and who you think approved the investment. Do you think that it was a profitable investment?

Chapter 3   Analysis of manufacturing systems

(3.1)   Using a simple domestic appliance as an example, discuss the way in which manufacturing systems might have developed from prototype to full production.

(3.2)   How would you characterize a specialist car manufacturer using the complexity uncertainty matrix?

(3.3)   What are the key nonfinancial performance measures of your business?

(3.4)   Prepare an IDEF diagram for a small department in a company with which you are familiar. Is the diagram comprehensive and acceptable to all relevant activities?

Chapter 4   Computer system fundamentals

(4.1)   Expand *Figure 4.4* and fill in more fields related to item and structure data, such as, for example, unit of measure, stock location, quantity on hand. Under what circumstances would engineering change data be part of the item record, or part of the structure record?

(4.2)   Expand *Figure 4.4* to include additional databases for suppliers, purchase orders, customers, sales orders, routing data and work centres. Which database would contain data about a customer who was also a supplier? Why are there no direct links between items and suppliers?

(4.3)   Specify the data that needs to be kept about shop orders. Why might data in the shop order database be different to the corresponding product structure and routing database?

(4.4)   List the advantages and disadvantages of magnetic stripe media compared with bar code printout for shop floor data recording.

(4.5)   Why are modems not needed for the transmittal of data on an ISDN?

(4.6)   What are the advantages and disadvantages of using standard software packages compared with having a system written by the company IS department?

Chapter 5   Information flow in manufacturing

(5.1)   Identify examples of the different kinds of items (parents, components, and products) for the following industries: engineering, chemical, oil, CPG (Consumer Packaged Goods), pharmaceutical, food, construction.

(5.2)   Discuss the areas of increased cost and complexity arising from changing a product's components when it is in the process of being manufactured; relate these to the information flow from design through development to process planning and manufacture.

(5.3)   From the business functions in the subsection 'Functional Information Requirements' and the 'information' descriptions in *Figure 5.4*, develop a matrix showing which function is responsible for originating and maintaining that information, and which functions have a need for access. Does this matrix help to emphasize the significance of some of those pieces of information and if so which?

(5.4) Discuss the benefits which could arise from the use of net change MRP compared with (*a*) a monthly planning run of MRP, and (*b*) order point control. Illustrate by reference to the product structure shown in *Figure 5.5*.

(5.5) Develop that part of the model in *Figure 5.4* that covers manufacturing to show a possible information flow on the shop floor of a company with which you are familiar. Particular attention should be paid to the flow between machines on the shop floor, and between the planning systems and the operational systems. Discuss the timing requirements of that flow in terms of minutes, hours, shifts, weeks etc.

Chapter 6   Simulation

(6.1) Tomorrow you are to be introduced to a brand new simulation system using a programming philosophy not so far revealed to the world. List the features and attributes that it must nevertheless possess in common with existing simulation systems.

(6.2) What do you consider are those features of a simulation system that make it (*a*) an effective way of describing the relevant parts of the real world, and (*b*) impressive to the people who are ultimately going to use it. Do these two lists coincide? Should they in an ideal world?

(6.3) 'Every simulation run should conclude its output with an explicit statement of the relevant costs of alternatives.' Discuss this statement, and the implication that in CIM all simulation is ultimately about money.

Chapter 7   Components of a CIM architecture

(7.1) What would be the impact upon manufacturing industry if the OSI Reference Model did not exist? How would computers from different manufacturers be connected together?

(7.2) Choose any one standard mentioned in Chapter 7 and investigate its exact current status, including which committees are responsible for it, what products on the market use it, and what standards are associated with it. Do not spend more than one month on this question.

(7.3) Why would a separate type of LAN be beneficial for the manufacturing shop floor?

(7.4) Select a convenient workstation. List its major components, and rate its power. What operating systems run on it, and what are the major applications it is used for? Can you find out where it is made, and estimate how long it will be before it is replaced by a significantly different model.

(7.5) Take a set of business decisions you are familiar with, and specify how an 'intelligent' computer system might handle them. Identify the facts that the computer would have to 'know'. Without any attempt to produce a program, try to define the rules by which the system might work.

(7.6)   At the level of the applications described in Chapter 2, draw up the ultimate CIM system for the company you work for, i.e. computerize everything not yet computerized, and link all existing computers (without regard to cost or business benefit). How many new computer systems are required, and how many different computers have to be connected together?

Chapter 8   Product and process design for CIM

(8.1)   Select an everyday item which can easily be disassembled and reassembled – a fountain pen would be a good example. Consider each separate component in turn and suggest possible changes in design to improve handlability and feedability. Now consider the assembly operations and make further design suggestions to allow easier automated or manual assembly. The approach taken can be based upon you own individual consideration of the design or upon the systematic techniques developed by Boothroyd.

(8.2)   From reading Chapter 8 and/or other sources of information:
• Describe how a point, line and circle can be specified using Cartesian coordinates. If you have access to a CAD system, describe how these geometric features are entered in that particular system.
• List some non-geometrical attributes of graphical elements.
• Describe the differences between wire frame, surface and solid models in CAD. Give examples where each type would find suitable application.
• In computer-aided drafting, what is the meaning of 'automatic dimensioning' and why can it be a misleading term?

(8.3)   Discuss the advantages of automated process planning compared with traditional manual systems. What are the differences between retrieval and generative process planning systems?

(8.4)   Select a sample machined item, for example part of a drive coupling or clutch plate assembly. Develop a sample classification and coding scheme which could be used to specify uniquely a family of parts of which the machined item would by a typical member.

Chapter 9   Planning and control in a CIM environment

(9.1)   For the company you work for, plot as a histogram the frequency of deliveries by supplier (e.g. hourly, twice a day, daily, twice a week, weekly etc.). What does this say about how close or far your company is from JIT production?

(9.2)   Under what circumstances would a Bill of Material have negative quantities in it?

(9.3)   Consider the manufacturing and assembly processes of a factory you are familiar with, and identify which of the capacity planning routines described in Chapter 9 are likely to be necessary.

(9.4) Low stock levels are a top priority of most manufacturing companies, a solution which is not necessarily the most economic but is at least measurable. Do you believe that CIM systems might develop such a precision of control of capacity and material flow that stock levels might, in a controlled manner, be allowed to increase again?

Chapter 10   Today's automation and intelligent manchines

(10.1)   Analyse why group technology has been so influential in the design of present-day factories. Identify the advantages and opportunities it provides, and the constraints it imposes.

(10.2)   Discuss, taking specific process examples, whether increasing sensory interactive behaviour or continuing mechanical engineering development will allow a larger variety of production processes to yield to a CIM approach.

Chapter 11   Customer/supplier communication

(11.1)   Consider a paint factory with three warehouses around the UK. Analyse whether a centralized or distributed system would be preferable by considering what data is needed by the warehouses and by the factory. Make the basic assumption that for the purposes of this exercise computer costs are the same for each solution. Does your answer change if the factory and warehouses are sited in the USA?

(11.2)   Consider the flow of information between a manufacturer of advanced technology products and a supplier of a component item, i.e. the supplier is likely to suggest design changes to the component and these changes may affect other components in the product. What data needs to pass between the two companies and how would engineering change be controlled? How would this communication be affected if different CAD systems were being used?

Index